A MANUAL FOR RESINS FOR SURFACE COATINGS

VOLUME I
(Second Edition)

Edited by **P. K. T. Oldring** BA, Ph.D
and **G. Hayward** C.Chem, MRSC

With contributions by a team of Senior Practising Chemists
within the Synthetic Resins, Paint, Printing Inks
and Coatings Industries in the UK

including

Roger Castle B.Sc
C. Standen B.Sc
G. Hayward C.Chem, MRSC

© 1987 Selective Industrial Training Associates Limited
London, United Kingdom

FIRST IMPRESSION 1986
SECOND IMPRESSION 1987
THIRD IMPRESSION 1991
FOURTH IMPRESSION 1993

Typesetting and Artwork produced by K-Dee Typesetters

Published by
SITA Technology
Gardiner House
Broomhill Road
London SW18 4JQ
England

ISBN 0.947798.04.8

INTRODUCTORY NOTE

This text book has been specifically designed for paint and ink technologists dealing with specific product related technical areas. Each aspect of the technology has been written by experts in that field. This has been done in order to provide comprehensive practical theory. In some cases this naturally involves repetition of theoretical aspects of the chemistry required for an in depth understanding of the particular technology under discussion.

A MANUAL FOR RESINS
FOR SURFACE COATINGS

CONTENTS

Chapter III — HARD RESINS, OLEORESINOUS MEDIA AND VARNISHES

Chapter I

An Introduction to Polymer Science

Chapter I

An Introduction to Polymer Science

In order to understand the following chapters of this book, it is necessary to have some basic knowledge of the principles of polymer science.

This chapter therefore, gives an introduction to the fundamentals of the structure of polymers (i.e. their constitution, chain structure, stereochemistry molecular weight distribution and glass transition temperature). It also gives an introduction to polymerisation theory.

The treatment of addition polymerisation is restricted to the free radical reactions of carbon-carbon double bonds. Condensation polymerisation theory is discussed in terms of the condensation reactions between acids and alcohols or amines.

THE STRUCTURE OF POLYMERS

THE CONSTITUTION OF POLYMERS

Polymers consist of numerous small units called monomer molecules, combined to form a large molecule. If the repeat unit is identical, then the polymer is termed a homopolymer and if there are more than one type of repeat units, it is known as a copolymer.

The properties of polymers are determined by the types of monomers from which they are constructed. This will be discussed in detail in subsequent chapters.

POLYMER CHAIN STRUCTURES

Monomer units (M) may be linked together to form chains of various types, e.g:

Linear   M — M — M — M — M — M — M — M

Branched   M — M — M — M — M — M — M — M
                          |
                          M
                          |
                          M

Cross-linked   M — M — M — M — M — M — M — M
                    | |
                    M   M
                    | |
                M — M — M — M — M — M — M — M
                    | |
                    M   M
                    | |
                M — M — M — M — M — M — M

The polymer structure has a major effect on the physical properties of the polymer. For example, linear structures are more soluble than cross-linked structures (for the same type of chemical repeat unit).

STEREOCHEMISTRY OF POLYMER CHAINS

There are two important topics relating to the actual micro-structure of polymer chains and these can be classified as:

  i) Combination of monomer unit

  ii) Tacticity and di-tacticity.

Combination of Monomer Units

The microstructure of the polymer depends on the orientation of the monomer unit within the chain.

For an addition polymer monomer unit: $CH_2 = CYX$ where X and Y represents any substituents, but $X \neq Y$ and Y can be H; there are three ways in which it can combine with a propagating (growing) chain where CYX is known as the head of a monomer unit and CH_2 the tail.

a) **Head to Tail (or Tail to Head)**

$$-CH_2 - {}^\bullet CYX + CH_2 = CXY \quad \rightarrow \quad CH_2 - CYX - CH_2 - {}^\bullet CXY$$

b) **Head to Head**

$$-CH_2 - {}^\bullet CYX + CYX = CH_2 \quad \rightarrow \quad CH_2 - CYX - CYX - {}^\bullet CH_2$$

Obviously the next monomer unit adds tail to head or tail to tail.

c) **Tail to Tail**

$$-CYX - \overset{\bullet}{C}H_2 + CH_2 = CYX \quad \rightarrow \quad -CYX - CH_2 - CH_2 - {}^\bullet CYX$$

The next monomer unit adds head to tail or head to head.

The major content of the polymer, depends upon the relative stabilities and energies of the monomer unit and radical, which are related to the nature of X and Y.

Generally, if the propagating species has the structure $-{}^\bullet CYX$, it is more stable than $-{}^\bullet CH_2$. This results in the fact that most polymers have a preponderance of a regular head to tail linkage of monomer units.

Examples include polystyrene, vinyl chloride, vinyl acetate, methyl vinyl ketone and vinyl alcohol.

Alkyl or halogen acrylates are believed to polymerise head to head, or tail to tail.

Obviously, there is always a proportion of all types of combination present in any high molecular weight polymer chain. With some monomers the proportion is appreciable, but with others, the probability of more than one type of combination is small enough to make it insignificant. As a general rule, ionic polymerisation will yield polymers with a higher degree of stereo regularity than free radical polymerisation of the same monomer entity.

The Tacticity of Polymers

A carbon atom in a polymer chain can have two different pendant groups attached to it, and since the lengths and structures of the pendant groups are different, the carbon atom can have two possible configurations. Such a carbon atom is termed asymmetric.

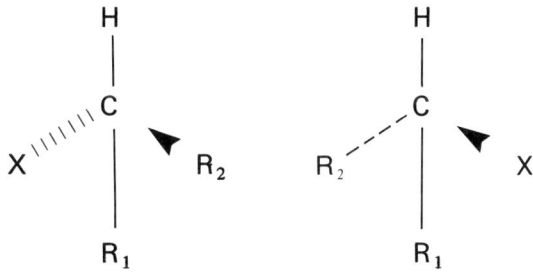

Where — denotes bonds in the plane of the paper and ← and |||| denote bonds at right angles to the plane of the paper.

Polymers with a highly regular distribution of both configurations, are known as tactic polymers. Those with random distribution are known as atactic polymers.

When there is a repetition of the same configuration of asymmetric carbon atoms along a polymer chain, the polymer is said to be isotactic. Where the configuration alternates at successive asymmetric carbon atoms, the polymer is said to be syndiotactic.

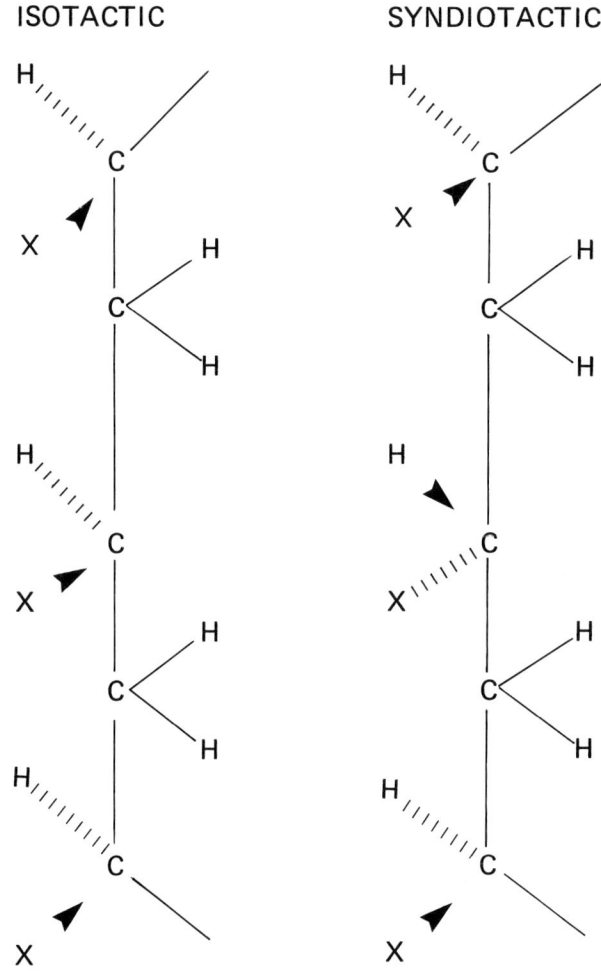

ISOTACTIC   SYNDIOTACTIC

Isotactic polymers can be represented as follows:

```
         H   Y   H   Y   H   Y
         |   |   |   |   |   |
      -  C - C - C - C - C - C -
         |   |   |   |   |   |
         H   X   H   X   H   X
```

Syndiotactic polymers can be represented as follows:

```
    H   X   H   Y   H   X   H   Y   H   X   H   Y
    |   |   |   |   |   |   |   |   |   |   |   |
    C - C - C - C - C - C - C - C - C - C - C - C -
    |   |   |   |   |   |   |   |   |   |   |   |
    H   Y   H   X   H   Y   H   X   H   Y   H   X
```

Where the asymmetric C atoms are consecutive, two types of di-isotactic structures can be formed as shown on page 8.

The di-syndiotactic analogues have not been isolated to date.

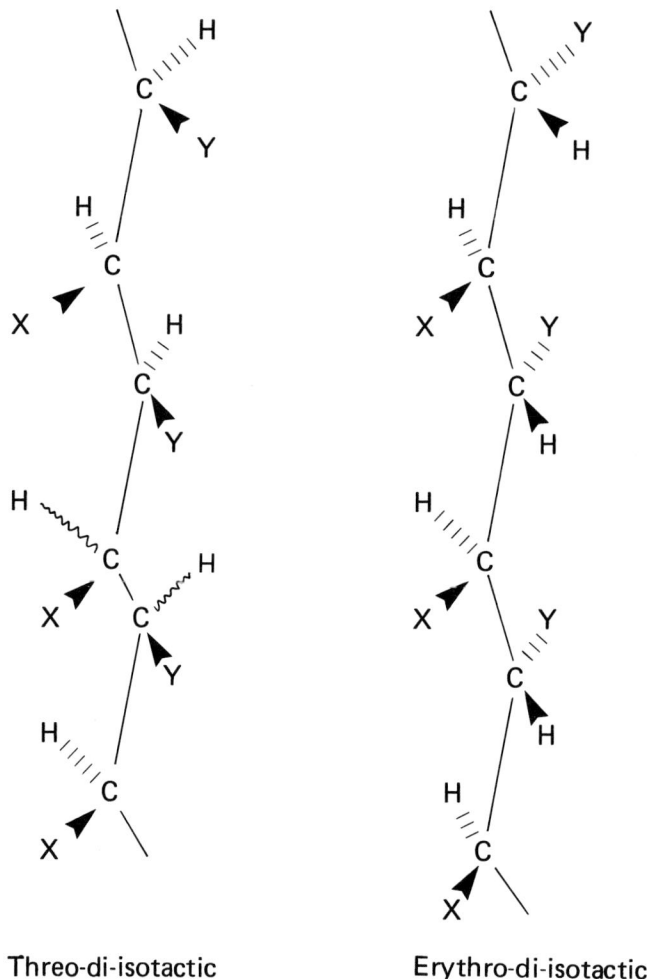

Threo-di-isotactic   Erythro-di-isotactic

Erythro-di-isotactic can be represented as follows:

$$
\begin{array}{ccccccccccc}
& X & Y & X & Y & X & Y & X & Y & X & Y \\
& | & | & | & | & | & | & | & | & | & | \\
-\!\! & C & - & C & - & C & - & C & - & C & - & C & - & C & - & C & - & C & - & C & - \\
& | & | & | & | & | & | & | & | & | & | \\
& H & H & H & H & H & H & H & H & H & H
\end{array}
$$

The alternative structure called threo-di-isotactic can be represented as:

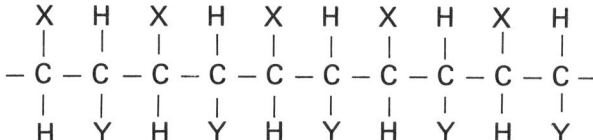

Meares (Polymers-Structure and Bulk Properties, published by Van Nosband 1967) discusses the probabilities of different types of tacticity and how to determine tacticity. Whilst, tacticity can effect solubility and melting point for many polymers (e.g. polypropylene) the significance for many commonly used polymers in the surface coating industry is generally unimportant because few homopolymers are used.

MOLECULAR WEIGHT

In the strictest sense the molecular weight of a polymer lacks a true definition, since the molecules are not all of the same size. Every polymer sample has a range of molecular weights, so that any molecular weight data for a given polymer, must represent something of an average value. Therefore, when characterising a polymer by molecular weight, it is usual to talk of molecular weight distribution, or a molecular weight average.

The picture is further complicated, because there are different ways of averaging the molecular weight of a polymer.

The average molecular weight of a polymer containing i repeat units (degree of polymerisation of i) and denoted Mi can be expressed in a number of ways.

The two most commonly used are:

a) **Number average molecular weight**

denoted $\bar{M}_n$ and expressed as

$$\bar{M}_n = \Sigma \, n_i M_i$$

where n_i is the number fraction of molecules of size i.

b) **Weight average molecular weight**

denoted $\bar{M}_w$ and expressed as

$$\bar{M}_w = \Sigma \, w_i M_i$$

where w_i is the weight fraction of molecules of size i.

The general equation for average molecular weight is:

$$\overline{M} = \Sigma\ NiMi^a\ /\ \Sigma\ NiMi^{(a-1)}$$

When a=1,M it is equal to the number average molecular weight and when a=2,M is the weight average molecular weight, viz:

(a = 1)
$$\overline{M}_n\ =\ \frac{\Sigma\ NiMi}{\Sigma\ Ni}\ =\ \frac{W}{N}$$

Where W=total weight of sample, and N=total number of molecules in sample.

(a = 2)
$$\overline{M}_w\ =\ \frac{\Sigma\ NiMi^2}{\Sigma\ NiMi}\ =\ \Sigma\ \frac{WiMi}{W}$$

Where Wi=weight fraction of molecules with degree of polymerisation i and W=total weight of sample.

Number average molecular weight is determined by measurement of the colligative properties of the polymer such as:
  i) Vapour pressure
  ii) Elevation of boiling point
  iii) Depression of freezing point
  iv) Osmotic pressure
  v) End group analysis.

Weight average molecular weight is determined by light scattering or sedimentation methods.
  Gel permeation chromatography measures hydrodynamic volumes of polymer molecules and different molecular weights can be calculated from this.

Viscosity average molecular weight: Viscosity determination of molecular weight is a method readily available for any resin chemist. Indeed viscosity control of a batch is frequently related to molecular weight. For any given solvent, polymer and temperature, viscosity increases with increasing molecular weight.
  For high molecular weights, the intrinsic viscosity η of a polymer in dilute solution can be related to molecular weight by the Mark Houwink equation:

$$\eta\ =\ k\ \overline{M}_v{}^a$$

Where k and a are constants for the particular solvent/polymer system and both are temperature dependent, k and a are calculated experimentally using polymers of known molecular weight as reference standards or obtained from published literature.

Normally the viscosity average molecular weight $\bar{M}_v$ lies between $\bar{M}_n$ and $\bar{M}_w$.

When $\qquad$ a $=$ 1 $\quad \bar{M}_v = \bar{M}_w$

$\qquad$ a $= -1$ $\quad \bar{M}_v = \bar{M}_n$

If a$=$O then $[\eta]$ is independent of molecular size. Since in practice a lies between 0.6 and 0.8 $\bar{M}_v$ lies between $\bar{M}_w$ and $\bar{M}_n$ but much closer to $\bar{M}_w$.

POLYDISPERSITY AND
MOLECULAR WEIGHT DISTRIBUTION

The relationship between $\bar{M}_n$ and $\bar{M}_w$ is a measure of the polydispersity of (D) of polymer.

Where $\qquad\qquad \dfrac{\bar{M}_w}{\bar{M}_n} \simeq 1$

the polymer has a narrow molecular weight distribution and said to be monodispersed. In practice polymers with a D value of 2−6 are considered as monodispersed systems.

Where $\qquad\qquad \dfrac{\bar{M}_w}{\bar{M}_n} > 1$

the polymer is considered to be polydispersed. The relationship can also be expressed through the diagrams.

a) Poussian Distribution

Frequency of
molecules of
a particular
molecular weight.

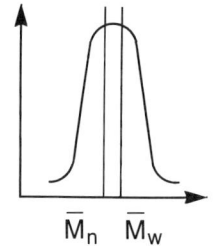

$\overline{M}_n$ $\overline{M}_w$

Narrow molecular
weight distribution

$$\frac{\overline{M}_w}{\overline{M}_n} \simeq 1$$

b) Gaussian Distribution

Frequency of
molecules of
a particular
molecular weight.

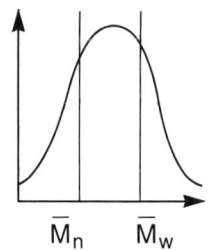

$\overline{M}_n$ $\overline{M}_w$

Random molecular
weight distribution

$$\frac{\overline{M}_w}{\overline{M}_n} > 1$$

c) Asymmetrical Distribution

Frequency of
molecules of
a particular
molecular weight.

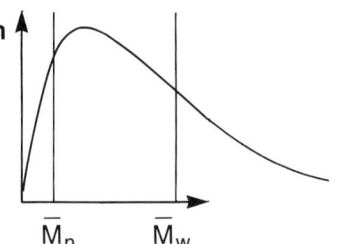

$\overline{M}_n$ $\overline{M}_w$

Broad moleculor
weight distribution

$$\frac{\overline{M}_w}{\overline{M}_n} \gg 1$$

GLASS TRANSITION TEMPERATURE

Simple substances can exist in three states, i.e. solid, liquid, vapour (or gas). The transition between the solid and liquid states is known as the melting point. The purer the substance the sharper (i.e. narrower range) the melting point.

Unlike simple substances, polymers consist of a mixture of molecules, all of differing molecular weights. For any given polymeric molecule, increasing its molecular weight increases its elastomeric nature, and the transition from solid to liquid is not as well defined as that of lower molecular weight material.

Furthermore, polymer chains can exist in crystalline or amorphous states. In their crystalline state, chains are in ordered, three-dimensional arrangements, whereas in the amorphous state they are in a tangled disordered array. Against this background it is possible for a single high molecular weight polymer chain to pass through several crystalline and amorphous regions, because a crystalline region is usually much shorter than the chain length. When there is a preponderance of crystalline regions (called crystallites), the polymer is considered to be crystalline.

For crystallinity to occur to an appreciable extent, it is necessary that some or all of the following conditions are met.

i) The polymer chains must be compact enough to come within close proximity of each other and to 'pack' together.

ii) The structure of the groups must be such that they can align themselves with groups in neighbouring polymer chains and form a close packed regular arrangement.

iii) Increasing polarity increases the tendency for crystalline behaviour because the molecular forces can hold the separate molecules together as if they were one.

The melting of a crystalline polymer is fairly well defined. The change from solid to liquid occurs over a small temperature range. The range, however, broadens with increasing molecular weight.

Amorphous polymers at low temperatures tend to be hard, transparent and glass like, hence the term glassy state. When the temperature is raised they either melt to a viscous liquid, or at higher molecular weights, they change from a glassy to a rubbery state. On raising the temperature further, there is a diffuse transition zone from rubber to viscous liquid. The temperature of transition from the glass to the rubbery state is called the glass transition temperature. It is normally represented as Tg.

For moderately crystalline polymers, Tg refers to the transition temperature at which a rigid crystalline polymer becomes a flexible crystalline polymer.

Frequently the Tg is defined as a second order transition, referring to a change in properties from a hard brittle glassy substance, to a soft flexible material.

It is measured by determining physical properties such as specific volume, refractive index, density and thermal conductivity at various temperatures, as shown on page 14.

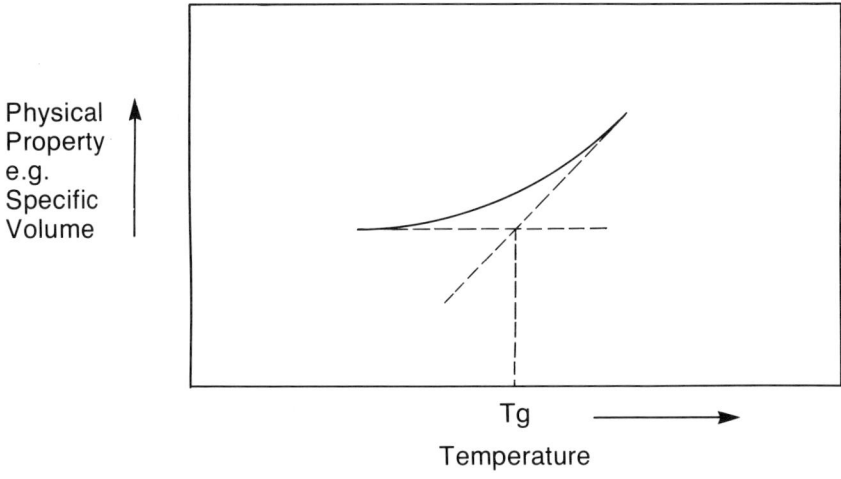

Temperature

There are two gradients and their intersection (when extrapolated if necessary) is the Tg.

A simple explanation of Tg is as follows. Molecular movements require thermal energy, and below the Tg there is insufficient energy available to promote molecular movement beyond the normal low energy vibrations of the chemical bonds.

Above the Tg there is sufficient energy for segments of the polymer chain to move. Further energy allows the co-operative movement of the molecular segments which results in flow. This is the viscous liquid state.

Whilst there is some degree of flow or co-operative movement in the rubbery state, the chain entanglements, etc., limit movement so that normally, externally applied forces are necessary for movement to occur (e.g. stretching a rubber band). If there are bulky substituents (side groups) in the carbon atoms of the main polymer chain (e.g. the phenyl group in polystyrene), it is possible to detect more than one Tg.

There are Tg's associated with the ability of the side chain to move, in addition to the movement of segments of the polymer molecule. Normally the Tg from the former is lower than that from the latter. Glass transitions are associated with amorphous regions of polymers, and the more amorphous a polymer, the greater the effect of the transition upon physical properties. The Tg's of the majority of resins used in the coatings industry lie between 0 and 60°C.

Why is Tg important to the resin chemist, and the surface coating and ink formulator? Many reasons can be given, but the major ones revolve around the required properties of the thermoplastic film.

Firstly, the coating must form a coherent film upon the evaporation of solvent or carrier liquid (dispersion medium). For latices this is referred to as Minimum Film Formation Temperature (MFFT). As a general rule the lower the Tg the more flexible and softer the film. Conversely the higher the Tg the harder, more rigid the film. If blocking is important, then the Tg of the coating must be considered and raised if necessary. Above its Tg a polymer will cold flow.

The two types of factors which influence the ease of rotation about single bonds can be considered as:

a) Factors which limit the intrinsic mobility of an isolated polymer chain, and
b) factors which limit the mobility of the polymer chain due to interactions with neighbouring polymer chains.

How can Tg be affected or modified? Tg is related to many factors.
  i) The ratio of polymer melting or softening point to Tg is roughly constant, because whatever affects Tg, also similarly affects melting point.
  ii) There is a relationship between Tg and many solution and film properties of a polymer. Burrel (Offic. Dig. Fed. Soc. Paint Technol (1962) 37 L, 131) discusses the relationship with solution viscosity, solvent release, drying speed, cure rate, flexibility, hardness, adhesion, impact resistance, tensile strength, abrasion resistance, colour and hiding power, permeability and electrical properties.

The Tg of a polymer can be affected by the following factors:
  i) Molecular weight
  ii) Degree of branching
  iii) Degree, if any, of cross-linking
  iv) Polarity of components
  v) Bulk of substituents
  vi) Plasticisation, (internal and external)
  vii) Entrained solvents
  viii) Tacticity
  ix) Copolymerisation
  x) Rigidity of backbone.

Molecular Weight

Some equations relating Tg to molecular weight have been derived. Most take the following form used by Fox and Flory.

$$Tg = Tg\infty - k/M$$

Where $Tg\infty$ is the Tg of an infinite molecular weight polymer, k is a constant and M is the number average molecular weight ($\overline{M}_n$). When M is large then plots of $1/Tg$ against $1/M$ are linear.

Degree of Branching

The free volume theory (Meares) can be used to understand how branching decreases Tg. A linear polymer molecule has two chain ends. Branching introduces more chain ends. If there are (w−2) branches then there are w chain ends.

The Tg is now:

$$Tg = Tg - kw/2M$$

Therefore, for any given molecular weight, increasing the branching decreases Tg.

Degree of Cross-linking

Intuitively, one would expect cross-linking to raise Tg because two or more chains are linked by an intermediate bridge which causes a reduction in the free volume, which in turn raises the Tg.

However, the degree and type of cross-linking is important. If the molar volume and structure of the cross-linking comonomer are markedly different to that of the monomer units of the polymer chain, then the network must be considered as a copolymer.

Polarity of Components

The more polar the components (either backbone or side groups) of the polymer chain, the more inter and intra molecular interactions (e.g. H bonding or Van der Waal forces) and the more restricted the molecular motion. Thus, the more energy required to overcome barriers to motion the higher the Tg of the polymer. If increasing polarity increases the degree of crystallinity, then other effects may obscure this rise. The melting point Tm increases however with increasing polarity, for a highly crystalline polymer.

Bulk of Substituents

Predicting the effect of altering the shape or size of substituents of the polymer is difficult because there are two counteracting considerations. Increasing the free volume decreases Tg, but decreasing the free rotation about the backbone carbon atoms by increasing the bulk of the substituents, increases Tg.

The use of long side chain comonomers (e.g. lauryl methacrylate) has a similar effect to branching. Since there are no more chain ends available the polymer chains are forced apart. This increases the free volume. The effect is larger than the restriction to free rotation, so that Tg is reduced.

However, if a bulky rigid side chain group is inserted (e.g. the phenyl group in poly-styrene replacing H in polyethylene) there is considerable resistance to rotation and the Tg is correspondingly increased.

Steric factors, due to replacing one group by a smaller or larger one, can cause the polymer chain to adopt a different configuration, thus affecting Tg.

Above a certain polymer chain length it is possible for the branching and 'forcing apart' effects to be outweighed by chain entanglements. These limit free rotation and hence increase Tg.

Plasticisation

Plasticisers are relatively low molecular weight substances of low volatility. When added to a polymer they act as a solvent and lower its Tg. This is because the free volume of a small molecule is much larger than that of a polymer at the same pressure and temperature. Typical plasticisers include di-butyl phthalate, di-isooctyl phthalate and tri-cresyl phosphate.

Even small amounts of plasticisers can lower Tg significantly. They have however to be compatible with the polymer, and in effect they 'dissolve' the polymer.

The problem with this type of plasticisation (which is called external plasticisation), is that frequently over a period of time, the plasticiser migrates from the film, embrittling it. This is not surprising when it is considered that plasticisers are relatively small mobile molecules, compared to the polymer, and their loss will raise the Tg. If the polymer is raised above ambient temperature and plasticiser migrates, then the film forms a glass.

The ideal characteristics of a plasticiser summarised by Bridson (Plastics Material published by Butterworths 1970) are:

1. The molecular weight should be about 300.
2. It should have a solubility parameter similar to that of the polymer.
3. If the polymer has a tendency to crystallise, the plasticiser should be capable of some specific interaction with the polymer.
4. It should not be a crystalline solid at ambient temperatures, unless it is capable of specific interaction with the polymer.

An alternative type of plasticisation is called internal plasticisation where a flexible molecule (e.g. adipic acid or 1,6 hexane diol), or a molecule giving a long side chain (e.g. lauryl methacrylate or ethylhexyl acrylate) are incorporated into the polymer chain to lower Tg.

Entrained Solvents

Any entrained solvent will act as a plasticiser until it evaporates from the film. Entrained solvent can be important for film formation. Small amounts of solvent can lower Tg markedly. For polyvinyl acetate emulsions, water is important for film formation which may occur at up to 10°C below the Tg of polyvinylacetate.

Tacticity

The tacticity of a polymer can affect its Tg. The more structured the polymer the higher the Tg. Thus, as a general rule atactic polymers have a lower Tg than either iso or syndio-tactic polymers formed from the same monomer units.

Copolymerisation

The Tg of a copolymer generally lies between the Tg's of homopolymers of its major monomers. Therefore it is a relatively easy matter to adjust the Tg of a copolymer by altering the ratio of 'hard' (high Tg) to 'soft' (low Tg) monomers, or by incorporating a harder or softer comonomer into the polymer. Under normal conditions the following approximation is adequate for calculating the approximate Tg of a copolymer.

$$\frac{1}{Tg} = \frac{W(a)}{Tg(a)} + \frac{W(b)}{Tg(b)} + \frac{W(c)}{Tg(c)} = \sum \frac{W(i)}{Tg(i)}$$

Where W(i) is the weight fraction of component (i) and Tg(i) is the Tg of the homopolymer of component (i). *Note* all temperatures are in degrees absolute (Kelvin). Mole fraction can be used in place of weight fraction, but the difference in the calculated Tg will be small.

Rigidity of Backbone

Obviously the backbone of a polymer is modified so that C−C or C−O bonds which are relatively flexible (i.e. rotation about the single bond can fairly easily occur) are replaced by fairly rigid groups like C=C or p-phenylene, then the Tg will be raised. Similarly, if a more flexible linkage like Si−O is incorporated then the Tg is reduced.

Another method of adjusting Tg by altering the backbone of the polymer is the incorporation of a more flexible unit (e.g. replacing ethylene glycol by 1, 6 hexanediol lowers the Tg of a polyester).

THE THEORY OF ADDITION POLYMERISATION

The type of addition polymerisation which is most used in the preparation of surface coating resins is the free radical initiated reaction of carbon-carbon double bonds.

We shall use this reaction, as an example, to give an overview of the theory of addition polymerisation, and to give an understanding of the principles involved. The major steps in any addition polymerisation are as follows:

　　1. Chain initiation.

　　2. Chain propagation.

　　3. Chain termination.

　　4. Chain transfer.

After these reactions have been considered, the kinetics of the addition polymerisation of homopolymers and copolymers will be discussed.

Initiation

Normally, free radicals are generated by the thermal decomposition of peroxides or azo compounds. Alternative initiation methods, e.g. Redox systems, are discussed in the chapter on vinyl polymerisation. One of the important concepts is the half life of the initiator.

The half life, denoted $t\frac{1}{2}$, is the time for half of the initiator to decompose at any particular temperature, and $t\frac{1}{2}$ decreases with increasing temperature, and varies significantly for different initiators and combinations of initiators.

The choice of initiator and processing temperature depends upon the system being polymerised, the method of initiator addition, and the time of addition of the monomer mixture.

To ensure complete reaction and minimal residual monomer at any stage of the process, it is necessary to have an excess of free radicals and not to deplete the system. There are tables of half lives of different initiators at different temperatures, and the method of selection is considered in more detail in the chapter Vinyl and Acrylic Polymers.

An initiator molecule I_2, can be considered to decompose to one or more free radicals $I^\bullet$ with rate constant k_d

$$I_2 \xrightarrow[\text{Heat}]{k_d} 2I^\bullet$$

Examples of typical thermal decomposition initiators are:

Benzoyl peroxide

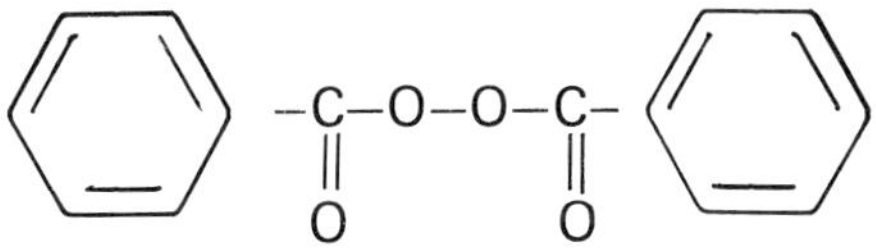

$$\text{Ph}-\overset{\displaystyle O}{\underset{\displaystyle \|}{C}}-O-O-\overset{\displaystyle O}{\underset{\displaystyle \|}{C}}-\text{Ph}$$

Azobis isobutyronitrile (AZDN)

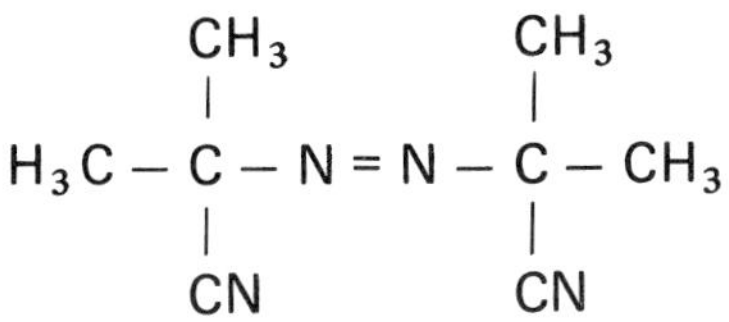

$$H_3C - \underset{\underset{CN}{|}}{\overset{\overset{CH_3}{|}}{C}} - N = N - \underset{\underset{CN}{|}}{\overset{\overset{CH_3}{|}}{C}} - CH_3$$

Depending upon the type of initiator, heterolytic or homolytic scission occurs when forming the free radicals. Some initiators have a greater tendency to cause radical induced decomposition than others and this is also reflected in the choice of initiator.

Only a percentage of these initiator radicals will initiate polymerisation, and the rate of initiation R_i is defined as the rate of formation of growing polymer chain radicals.

$$R_i \quad = \quad 2\,f\,k_d\,[I]$$

Where [I] is the molar concentration of the initiator, and f is the initiator efficiency (which is the fraction of radicals which initiate polymerisation).

$I^\bullet$ can be considered as a primary radical which is capable of reacting with a monomer molecule (M) to give a polymer chain initiating radical.

$$I^\bullet + M \longrightarrow IM^\bullet$$

i.e.

$$I^\bullet + CH_2 = \underset{\underset{Y}{|}}{CH} \longrightarrow I - CH_2 - \underset{\underset{Y}{|}}{CH^\bullet}$$

Alternatively, the free radicals can recombine:

$$I^\bullet + I^\bullet \longrightarrow I_2$$

Chain Propagation

One of the most important reaction mechanisms is chain propagation, during which monomer units are combined to form polymers. The chain radical can add other monomer molecules to it, and still retain the free radical reactivity.

$$IM^\bullet + M \longrightarrow IMM^\bullet$$

$$I(M)_n\,M^\bullet + M \xrightarrow{\ \ k_p\ \ } I(M)_{n+1}\,M^\bullet$$

k_p is the rate constant for polymerisation or chain propagation. Competing non-chain propagating reactions are also occurring. A reaction can only be considered to be chain propagating if it obeys the following conditions:

i) The overall molecular weight of the resulting radical is increased.

ii) The resulting radical remains capable of further molecular weight increasing reactions.

The propagation rate (R_p) depends upon the concentration of growing chains and monomer. But this can be considered to be the rate of monomer consumption and so it is the overall rate of polymerisation.

$$R_p \quad = \quad k_p \, [M] \, [IM_n \, M^{\bullet}]$$

Termination

There are several methods by which propagating chain radicals are terminated. Two important ones can be summarised as follows:

i) **Combination:** This is the simplest method of termination and involves the combination of two propagating chain radicals to form one species with rate constant k_{tc} thus:

$$IM_n \, M^{\bullet} + IM_m \, M^{\bullet} \quad \xrightarrow{k_{tc}} \quad IM_{(m+n+2)} \, I$$

ii) **Disproportionation:** This involves the collision of two propagating species to extinguish both radicals and leave two molecules with rate constant k_{td}, thus:

$$IM_n - \underset{\underset{H}{|}}{\overset{\overset{H}{|}}{C}} - \underset{\underset{H}{|}}{\overset{\overset{Y}{|}}{C^{\bullet}}} \quad + \quad IM_m - \underset{\underset{H}{|}}{\overset{\overset{H}{|}}{C}} - \underset{\underset{H}{|}}{\overset{\overset{Y}{|}}{C^{\bullet}}}$$

$$\downarrow k_{td}$$

$$IM_n - CH = CHY + IM_m - CH_2 - CH_2 Y$$

In effect an H atom has moved to give saturated and unsaturated compounds. The double bond is available for further reaction, but because of the reduced mobility of the molecule due to its large molecular size (c.f. monomer molecule), the probability of further reaction is much reduced.

Polystyrene chains terminate mostly by combination, unlike polymethyl methacrylate chains which tend to terminate by disproportionation above 60°C. Other termination reactions, which can be defined as removing the 'radical' of the propagating species, including the following:

— collision with the walls or stirrer of the reaction vessel
— reaction with contaminant
— reaction with oxygen (in air)
— reaction with free radical scavengers.

Chain Transfer

There is also another type of reaction which terminates the growing chain, but transfers the radical to a new species. This is termed chain transfer. The commonest types of this reaction are:

i) Chain transfer to solvent.

ii) Chain transfer to a substance deliberately added to control molecular weight. This is termed a chain transfer agent.

iii) Chain transfer to a species which forms a stable radical (e.g. quinones). These are called inhibitors and this process should be truly classified as termination.

iv) Chain transfer to monomer or oligomer, or polymer.

v) Transfer from propagating chain to initiator.

Many polymerisations take place in solution and (depending upon the nature and amount of solvent present), a certain degree of transfer to solvent occurs.

Lists of chain transfer constants are available in many textbooks. The reaction obviously reduces molecular weight and can be summarised as:

$$IM_m M^{\bullet} + SX \xrightarrow{\quad k_{ts} \quad} IM_m MX + S^{\bullet}$$

where k_{ts} is the rate constant and S represents solvent residue, and X is normally H but can be a halogen (particularly where carbon tetra chloride is used as a transfer agent). The radical $S^{\bullet}$ is available for further reaction with another monomer molecule to initiate a propagating chain or other free radical reactions.

As can be seen from this reaction scheme, the overall concentration of free radicals is not reduced. This typifies transfer reactions.

The amount of transfer to solvent is generally insufficient to produce relatively low molecular weight polymers and specific substances termed chain transfer agents are added when low molecular weight polymers are required.

$$IM_n M^\bullet + XH \xrightarrow{k_{tm}} IM_{(n+1)} H + X^\bullet$$

where XH is the chain transfer agent and k_{tm} is the rate of chain transfer. $X^\bullet$ is available for further free radical reaction.

Chain transfer agents are often present, up to $1-2\%$, but as the commonly used ones are mercaptans (e.g. tertiary dodecyl mercaptan), the levels are kept as low as possible to minimise odour.

Transfer to monomer can result in either an unsaturated monomer radical and saturated polymer, or polymer with terminal unsaturation and a saturated monomer radical, viz:

$$IM_n M^\bullet + CH_2 = CHY \xrightarrow{k_{tm}} IM_{(n+1)} H + CH_2 = C^\bullet Y$$

or

$$IM_n \, CH_2 - \underset{\underset{X}{|}}{\overset{\overset{H}{|}}{C}}{}^\bullet + CH_2 = CHY \longrightarrow IM_n \, CH = CHX + CH_3 - \underset{\underset{Y}{|}}{\overset{\overset{H}{|}}{C}}{}^\bullet$$

The monomer species is then capable of chain propagation.

It is possible for initiator to undergo chain transfer producing a low molecular weight initiator molecule and a polymer propagating species.

$$I^\bullet + CH_2 = CHY \longrightarrow IH + CH_2 = C^\bullet Y$$

It is also possible for a propagating chain to transfer its radical character to an initiator molecule, thus:

$$IM_n M^\bullet + I_2 \longrightarrow IM_n MI + I^\bullet$$

The initiator radical is then available for further reaction.

THE KINETICS OF
ADDITION POLYMERISATION

Having defined individual reactions, it is instructive to consider the rate of polymer formation, using the approach of Flory (Principles of Polymer Chemistry, Cornell University Press 1967).

As already discussed, the rates of initiation and propagation are:

$$R_i = 2fk_d [I]$$

$$R_p = kp [M] [IM_n M^\bullet]$$

Termination is a more complicated situation, because many more types of competing reactions are occurring.

The majority of termination reactions are either by combination or disproportionation. Both involve the bi-molecular reaction of two propagating chains, viz:

$$IM_n M^\bullet \quad + \quad IM_m M^\bullet \xrightarrow{\ \ k_t\ \ } \text{non-radical species}$$

If the rate constant is denoted k_t (i.e. termination) then the rate of reaction, independent of which type, depends upon the disappearance of the propagating chain as follows:

$$R_t = -d [M^\bullet]/dt = 2k_t [IM_m M^\bullet] [IM_n M^\bullet]$$

The Factor 2 is a result of the disappearance of two radicals per termination reaction.

For convenience, if $IM_m M^\bullet$ is assumed to be equivalent to $IM_n M^\bullet$, and all propagating species can now be written as $M^\bullet$.

Then $$R_t = 2k_t [M^\bullet]^2 \quad \text{and} \quad Rp = k_p [M] [M^\bullet]$$

A steady state is normally attained fairly rapidly, with the concentration of free radicals being constant overall.

Thus, the rate of formation of free radicals equals the rate of their disappearance. Therefore:

$$Ri = R_t$$

$$2fk_d [I] = 2k_t [M^{\bullet}]^2$$

$$[M^{\bullet}] = \left(\frac{fk_d [I]}{k_t} \right)^{\frac{1}{2}}$$

Substituting for $[M^{\bullet}]$ in the propagation reaction:

$$Rp = kp [M] \left(\frac{fk_d [I]}{k_t} \right)^{\frac{1}{2}}$$

This indicates that the rate of reaction is proportional to the square root of the initiator concentration, and to the monomer concentration. Increasing temperature increases k_d k_p and k_t, and in practice a 10°C rise in temperature doubles the rate of conversion, but decreases molecular weight.

Under normal circumstances, the initiator efficiency f, approaches unity but if it does not and is very low, then Rp depends mainly on monomer concentration. A similar expression can be used for radiation initiated systems and here efficiency is important.

Plots of Rp against [I] indicate initiator efficiencies. Flory discusses initiator efficiency in great detail and gives many examples. He also discusses problems occurring with different initiator breakdown mechanisms. If f appears to be greater than unity, then an auto-catalytic mechanism may be occurring.

The average kinetic chain length (v) can be defined as the number of monomer units consumed, per active centre formed, and can be related to the rates of propagation and initiation which, under steady conditions, equals the rate of termination.

$$v = Rp / Ri = R_p/R_t = \left(\frac{kp}{2k_t} \right) \frac{[M]}{[M^{\bullet}]}$$

Provided the propagation and termination processes are second order then the kinetic chain length is inversely proportional to the concentration of radicals, and hence on the rate of polymerisation. Substituting for $[M^\bullet]$ in $Rp = kp [M] [M^\bullet]$ gives:

$$v = \frac{(kp\ [M])^2}{2k_t\ Rp}$$

Eliminating $[M^\bullet]$ as before gives for an initiated polymerisation:

$$v = \frac{kp\ [M]}{2\ (fk_d\ k_t\ [I])^{\frac{1}{2}}}$$

Whilst an increase in the concentration of the initiator increases the rate of polymerisation, it decreases the molecular weight.

For thermal polymerisation (as against initiated polymerisation), the initiation is second order and kinetic chain length is independent of monomer concentration with:

$$v = \frac{kp}{2\ (k_i\ k_t)^{\frac{1}{2}}}$$

The number average degree of polymerisation, $\bar{x}_n$, is the average number of monomer units per polymer chain, and can be considered as:

$$\bar{x}_n = \frac{\text{rate of propagation}}{\text{combined rate of all termination reactions}}$$

If termination is by disproportionation $\bar{x} = v$. If however, it is by combination $\bar{x} = 2v$.

Some of these principles can be illustrated by considering the simplest case of chain transfer:

$$M^\bullet + SH \quad \xrightarrow{k_s} \quad MH + S^\bullet$$

$$R_s = k_s [M^\bullet] [SH]$$

$$\bar{x} = \frac{Rp}{R_t + R_s} = \frac{Rp}{R_i + R_s}$$

$$\frac{1}{\bar{x}} = v + \frac{Rs}{Rp} = v + \frac{k_s [SH]}{kp [M]}$$

This means that the greater the transfer rate constant, and the greater the concentration of transfer agent, the lower the molecular weight.

Some important conclusions can be drawn from the preceding kinetic study (Brydson).

1. The formation of a polymer molecule takes place instantaneously once an active centre is formed. This means that either polymer or monomer exist at any instant in time, apart from a very small number of propagating chains. This is in complete contrast to condensation polymerisation as we shall see later.

2. Increasing reaction time only increases the degree of conversion and not the molecular weight. There are however, exceptional circumstances, where grafting or other effects may occur to increase molecular weight, when 100% conversion of monomer to polymer is approached.

3. Increasing temperature or initiator concentration increases the rate of conversion, but decreases molecular weight.

4. Transfer reactions reduce the degree of polymerisation without affecting the rate of conversion.

5. The molecular weights of the polymers formed are statistically distributed. Molecular weight distribution is affected by the technique of polymerisation. If all the monomer is present initially, then a polymer with a different molecular weight distribution is obtained to one in which monomer is continuously added during processing.

Auto acceleration commonly called the Tromsdorf effect can occur. In essence the molecular weight increases rapidly as does the viscosity. Monomer units are rapidly added to a growing chain, giving an exothermic and potentially dangerous situation. The free radicals can be considered to be active, but trapped due to the inability of the reaction mixture to be thoroughly mixed, which is a consequence of the onset of the Tromsdorf effect. Some monomers are more prone to this effect than others and methyl methacrylate is probably the commonest. The onset normally occurs at degrees of conversion in excess of 60% (monomer to polymer) and control of the reaction condition is difficult. A runaway exothermic reaction is likely to occur.

COPOLYMERISATION

For the majority of surface coating applications a polymer composed of identical repeat units does not perform adequately. A mixture of two or more types of monomer molecules can be polymerised in the presence of each other by a free radical mechanism, to give a polymer containing several different monomer moieties.

The properties of the resulting polymer are normally significantly different to those of a blend of the homopolymers of the different monomers. For example, if an insoluble homopolymer is blended with a soluble homopolymer the insoluble homopolymer is still insoluble. If however, a mixture of the two monomer types are copolymerised, then the resulting copolymer may be soluble.

Copolymers consisting of two monomer moieties can be constructed in various ways. If A and B represent the monomer moieties, then the following are some of the different types of copolymers which can be formed.

AABABAAABBABBBBAB — random

ABABABABABABAB   — alternating

AAAAABBBBBBBAAAAA — block

AAAAAAAAAAAAAAAA — graft
BBBB   BBBB

THE KINETICS OF COPOLYMERISATION

In order to construct copolymers with known compositions, it is important to understand the kinetics of copolymerisation. Restricting the discussion to the chain propagating reaction with only two monomer moieties M_1 and M_2, the following reactions can occur with rate constants k_{ij}.

$$M_1{}^\bullet + M_1 \xrightarrow{k_{1,1}} M_1 M_1{}^\bullet$$

$$M_1{}^\bullet + M_2 \xrightarrow{k_{1,2}} M_1 M_2{}^\bullet$$

$$M_2{}^\bullet + M_2 \xrightarrow{k_{2,2}} M_2 M_2{}^\bullet$$

$$M_2{}^\bullet + M_1 \xrightarrow{k_{2,1}} M_2 M_1{}^\bullet$$

Consider the steady state where the rate of appearance equals the rate of disappearance for free radicals. If the chains are long, initiation and termination reactions are negligible by comparison. Using the treatment outlined by Flory the steady state condition is approximated by:

$$k_{2,1} [M_2{}^\bullet] [M_1] = k_{1,2} [M_1{}^\bullet] [M_2]$$

The rates of monomer consumption are given by the following:

$$-d [M_1]/dt = k_{1,1} [M_1{}^\bullet] [M_1] + k_{2,1} [M_2{}^\bullet] [M_1]$$

$$-d [M_2]/dt = k_{2,2} [M_2{}^\bullet] [M_2] + k_{1,2} [M_1{}^\bullet] [M_2]$$

It is possible to rearrange this by eliminating a radical concentration and dividing the last two equations to obtain the following.

$$\frac{d [M_1]}{d [M_2]} = \left(\frac{[M_1]}{[M_2]}\right) \left(\frac{r_1 [M_1] / [M_2] + 1}{[M_1] / [M_2] + r_2}\right)$$

where r_1 and r_2 are monomer reactivity ratios defined as follows:

$$r_1 = k_{1,1} / k_{1,2}$$

and

$$r_2 = k_{2,2} / k_{2,1}$$

These ratios can be used to predict the composition of the resulting polymer formed from two monomers. The ratio r_i can be considered as the tendency of monomer M_i to react with itself (i.e. homopolymerise) or with the other monomer (i.e. copolymerise).

A value of $r_1 = 1$ indicates that $k_{1,1}$ and $k_{1,2}$ are of similar magnitude and there is an even chance of M_1 or M_2 adding to the propagating chain. If $r_1 = 1/r_2$ (or $r_1 r_2 = 1$) then the two radicals show the same preference for adding the same or the other monomer.

The copolymer is classifed as an ideal one, because the monomer units are arranged completely at random along the chain, independent of feedstock ratios. This is as expected because $k_{1,1}/k_{1,2} = k_{2,1}/k_{2,2}$.

Where r_1 is significantly greater than unity, the propagating species only adds to its own monomer units (i.e. M_i), resulting essentially in homopolymerisation, particularly if the other reactivity approaches zero. For the system styrene, vinyl acetate, r_1 and r_2 are 55 and 0.01 respectively, and very little copolymer is formed.

When r_i is significantly less than unity, the propagating species prefers to add the other monomer units. With styrene and methyl methacrylate monomer combinations r_1 and r_2 are 0.52 and 0.46 respectively. This means that there is twice the probability of the opposite monomer being added than the same one. If the product $r_1 r_2$ approaches zero (or if $r_1 = r_2 = O$), then the opposite monomer is nearly always added, thus forming an alternating copolymer.

Most reactivity ratios are essentially independent of temperature, solvents, and other factors, which affect the overall kinetic rate, because generally all four rates ($k_{1,1}$ $k_{1,2}$ $k_{2,2}$ $k_{2,1}$) are similarly affected. However, the method of initiation has a significant effect. There are large differences between ionic and free radical polymerisation. Indeed, it is only possible to obtain certain types of copolymers by using a specific initiation system. Pepper has published much data relating to the differences in reactivity ratios, between ionic and free radical polymerisations. Examples of the use of reactivity ratios are given in the vinyl and acrylic chapter.

Most addition polymers used in the surface coating industry are formed by free radical polymerisation. This is partly because ionic polymerisation conditions and initiators are often less compatible with the environments encountered in the processing of surface coating polymers, than the more durable free radical ones.

The values of r_1 and r_2 may be determined experimentally by analysing the copolymer to find the ratio of combined monomers. In practice, it is often possible to find the values in published lists.

There is an alternative approach to determining experimentally the values of r_1 and r_2 for every mixture of monomers 1 and 2. It involves calculating values from experimentally determined other values. Alfrey and Price developed an empirical expression relating monomer reactivity, and an electrostatic contribution which relates to the tendency of any monomer to form alternating copolymers. The expression generally known as Q and e values is of the following form:

$$k_{1,1} = P_1 Q_1 \exp(-e_1{}^2)$$

$$k_{1,2} = P_1 Q_2 \exp(-e_1 e_2)$$

thus $$r_1 = (Q_1/Q_2) \exp[-e_1(e_1 - e_2)]$$

and similarly

$$r_2 = (Q_2/Q_1) \exp[-e_2(e_2 - e_1)]$$

thus $$r_1 r_2 = \exp[-(e_1 - e_2)^2]$$

The reciprocal of the monomer reactivity ratio $1/r_i$ can be considered as the monomer reactivity which represents the relative reactivity of like and unlike monomers with the propagating species. It is known that the structure of the monomer influences its reactivity, e.g. a phenyl group adjacent to a double bond increases the reactivity more than that for a methyl group, and this can be related to degree of resonance stabilisation (or delocalisation of electrons). This contribution to overall reactivity is represented by P_i and Q_j where P_i relates to the reactivity of radical $M_1{}^\bullet$ and Q_j to monomer M_j. The other factor is due to the polarities of the double bonds in the two monomers and e_1 and e_2 can be considered to be proportional to the residual electrostatic charges in the respective reacting groups. To obtain the above equations it has been assumed that e_i is the same for both the monomer M_i and the radical $M_i{}^\bullet$. This assumption is generally adequate because it depends upon the substituent and its polarisation of the double bond.

If ε_1 and ε_2 represent charges, D the di-electric constant, and k the Boltzmann constant, then $e_1 e_2 = \varepsilon_1 \varepsilon_2 / DkTx^*$ where x^* is the distance separating the charges in the activated complex.

Q and e values are attributed to individual monomers depending upon experimentally determined values of reactivity ratios, and these Q and e values then allow reactivity ratios to be calculated for any pair of monomers, for which there are Q and e values available. Whilst the results are approximate, they are normally more than adequate for most situations encountered in the surface coating area. Q and e values for individual monomers are normally taken from published lists.

The unreacted monomer ratio changes, as polymerisation proceeds, and this will cause the instantaneous composition of the polymer to vary. It is possible to calculate feed ratios of monomers to obtain a given composition, provided r_1 and r_2 are known. The basic equation is:

$$F_1 = (r_1 f_1{}^2 + f_1 f_2) / (r_1 f_1{}^2 + 2f_1 f_2 + r_2 f_2{}^2)$$

Where F_1 is the mole fraction of monomer M_1 in the increment of copolymer formed at a given stage in the copolymerisation, and f_1 is the mole fraction of unreacted monomer M_1 in the feed.

$$(f_1 = 1 - f_2)$$

As a general rule $F_1 \neq f_1$, and both change as polymerisation proceeds, and the polymer obtained over a finite range of conversion will consist of the summation of increments of polymer differing progressively in mole fraction F_1.

When $r_1 = 1/r_2$ the equation is simplified to:

$$F_1 = r_1 f_1 (r_1 f_1 + f_2)$$

The composition of the polymer at any instant can be approximated by the following expression, where m_1 and m_2 are the proportions of monomers M_1 and M_2 in the polymer.

$$\frac{m_1}{m_2} = \frac{d[M_1]}{d[M_2]} = \frac{[M_1]}{[M_2]} \left(\frac{r_1 [M_1] + [M_2]}{r_2 [M_2] + [M_1]} \right)$$

THE THEORY OF CONDENSATION POLYMERISATION

Condensation polymers are formed from the reaction of monomers which contain two or more reactive groups. For example, consider the formation of a linear polymer molecule by the condensation reaction between functional groups A and B, which are either contained within the same monomer $A-B$ (e.g. a hydroxy acid), or in different molecules $A-A$ and $B-B$ (e.g. adipic acid and ethylene di-amine).

$$n\ A-B \rightarrow A-B-A-B-A-B-A-B-\text{etc};$$

$$n\ A-A + n\ B-B \rightarrow A-A-B-B-A-A-B-B \text{ etc.}$$

The reactivity of the functional group is effectively independent of chain length, and the chain does not only grow by the addition of a monomer molecule, it also grows by the condensation of oligomers.

$$A-B-A-B + A-B-A-B-A-B$$

$$\rightarrow A-B-A-B-A-B-A-B-A-B$$

As the process proceeds, the molecular weight distribution broadens. In contrast to addition polymerisation, high molecular weights are not formed till the process is almost completed. The kinetics which are used for simple molecules can be successfully applied to condensation polymerisation if the rate expressions are written in terms of functional group concentrations, rather than mole fractions. Condensation polymerisation involves the reactions of two molecules to form a new molecular species, usually with the elimination of a smaller molecule such as water.

It has been shown that the reactivity of a functional group in a condensation polymer is effectively independent of the chain length and any other functional groups present, provided they are sufficiently far enough apart. The kinetics which are used for simple molecules can be equally successfully used for understanding the reactions occurring during polymerisation. Furthermore, the rate expressions can be written in terms of participating functional groups per unit volume rather than mole fraction.

For convenience, polyesterification will be considered, but these principles apply to most condensation polymers. The basic esterification reaction is:

$$R-CO_2H + HO-R^1 \rightarrow RCO_2R^1 + H_2O$$

The order of this reaction depends upon the presence or absence of a strong acid. In the absence of a strong acid a second molecule of acid acts as a catalyst, thus giving the following rate expression for the disappearance of acid with respect to time.

$$-d\,[CO_2H]\,/\,dt \;=\; k\,[CO_2H]^2\,[OH]$$

Using the approach of Flory (Principles of Polymer Chemistry), the fraction of the functional groups which have reacted at a time interval (t), to those initially present, can be defined as the extent of reaction (p). If C_0 represents the initial concentration of functional groups, and the concentrations are all equal at time t and are denoted C then:

$$2\,kt \;=\; 1/C^2 - Constant$$

This is the standard integrated form of a third order reaction equation. The constant is $1/C_0^2$ if the reaction rate is constant throughout the process.

$$R-COOH + HO-R^1 \rightarrow R-COO-R^1 + H_2O$$

Initial concentration

C_0	C_0	nil	nil

Concentration after time t

C	C	pC_0	pC_0

If p groups have reacted (where p represents the extent of reaction) the equation can be written as:

$$(1-p)C_o \qquad (1-p)C_o \qquad pC_o \qquad pC_o$$

where $C=(1-p)C_o$.

Thus,

$$2C_o^2 \, kt = \frac{1}{(1-p)^2} - \text{Constant}$$

(Constant is 1 if equal concentrations are present initially)

A plot of $\frac{1}{(1-p)^2}$ against t should be linear. Flory obtained such relationships for di-basic acids and glycols (e.g. adipic acid and ethylene glycol). Consider a linear condensation polymerisation of a dibasic acid and a glycol.

Provided there are no side reactions, i.e. every reaction involves addition to the structural unit, then the number average degree of polymerisation $\bar{x}_n$ is defined as:

$$\bar{x}_n = \frac{\text{No. of units}}{\text{No. of molecules}} = \frac{1}{(1-p)}$$

$$\therefore \bar{M}_n = \frac{M_o}{1-p}$$

where a structural unit is defined as the residue from a glycol and dibasic acid, and the number of structural units equals the total number of bi-functional monomers initially employed. The number of unreacted carboxyl groups equals the number of molecules present. The number of molecules per unit volume is $(1-p) \, C_o$. The mean molecular weight per monomer unit is M_o.

An acid catalysed reaction is second order, and provided the concentration of the catalyst, which is small and constant, is contained in the rate constant k^1, the following equation holds:

$$C_o \, k^1 \, t = \frac{1}{1-p} - \text{Constant}$$

For equal concentrations the constant is unity and plots of $1/(1-p)$ against t should be linear. Flory has shown this to be the case for ethylene glycol, adipic acid catalysed by 0.4 mole % of p-toluene sulphonic acid.

The extent of reaction p also represents the probability that a functional group has reacted at time t, with the probability of finding an unreacted group being $(1-p)$ and an n-mer will contain $(n-1)$ reacted functional groups. The functional groups behave independently of each other, and the probability of any functional group entering into reaction is also an independent probability event (which is p per reaction). Thus, the probability of $(n-1)$ ester linkages being formed is the product of each probability, which is $p^{(n-1)}$ and there is also the probability that the molecule contains at least n units. To limit the chain to n units, the nth carboxyl group has to be unreacted and the probability of this is $(1-p)$. Therefore, the probability of finding an n-mer has been shown by Flory to be $p^{(n-1)} \cdot (1-p)$. This is also equal to the fraction of all the molecules which are n-mers.

Assume equivalent concentrations of functional groups (or self condensation) and only linear reactions.

Let N represent the total number of molecules and N_n the total number of n-mers. Then,

$$N_n = N \, p^{(n-1)} \, (1-p)$$

If the total number of structural units is N_o, then

$$N = N_o \, (1-p)$$

Thus,

$$N_n = N_o \, (1-p)^2 \, p^{(n-1)}$$

Ignore the weight of the end groups which is negligible for fairly large molecular weights, then the weight fraction of n-mers denoted W_n is given as follows:

$$W_n = nN_n / N_o = n(1-p)^2 \, p^{(n-1)}$$

Consider now plots of w_n and N_n against n for different extents of reaction (i.e. different values of p which have been reproduced from Flory's book and Flory (J Chem Phys 12,425, 1944).

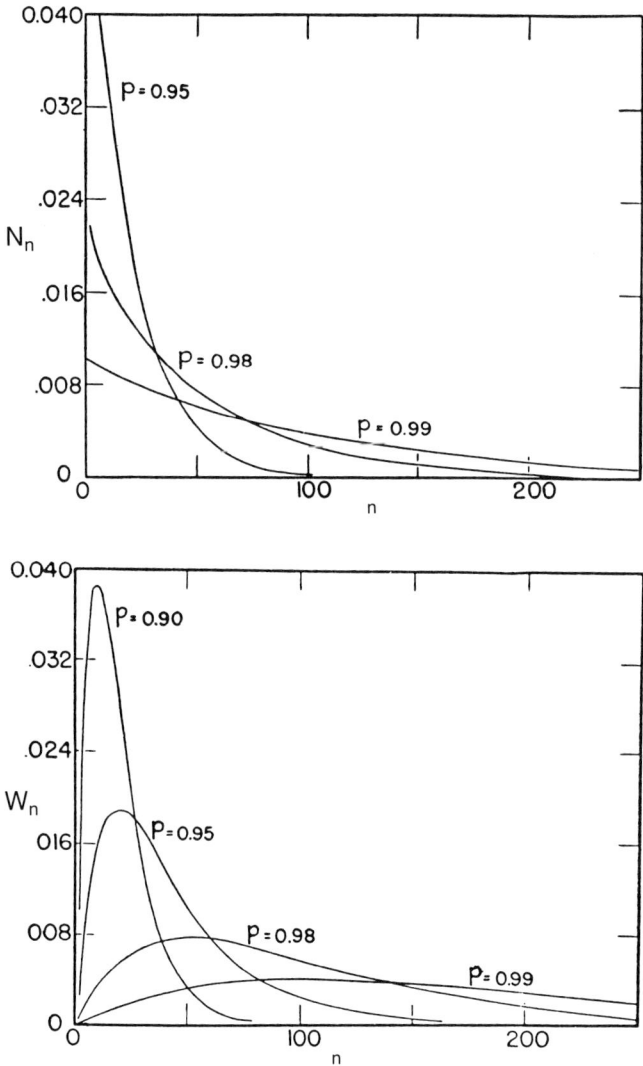

This series of curves illustrates some of the major differences between condensation and addition polymerisation. With condensation polymerisation, high molecular weights are only obtained at high degrees of conversion. Reaction proceeds stepwise with the molecular weight distribution broadening during processing. On a number basis there are more monomers present than any other species irrespective of the degree of polymerisation, but on a weight average they are insignificant with increasing p.

Maxima in the W_n *vs* n curve occurs near the number average value (i.e. $1/(1-p)$). If stoichiometric quantities of monomers are initially present, then the average degree of polymerisation n is given by the following expression:

$$\bar{x}_n \ = \ \frac{1}{1-p}$$

when all of a functional group species has reacted, then $p=1$.

From this expression it can be seen that very high molecular weights will not be produced till p is almost 1. For example, when $p=0.95$ and the reaction is 95% complete, the degree of polymerisation is only 20, i.e. $(1/(1-0.95))$, and the molecular weight average is therefore only 20 times the molecular weight of the monomers. At 99.5% reaction, the degree of polymerisation has risen to 200, i.e.:

$$\left(\frac{1}{(1-0.995)} \right)$$

However, unlike addition polymerisation, the condensation reaction is reversible as the equations used in these derivations indicate. It is possible to reduce the molecular weight of a condensation polymer by adding water, or (depending upon the concentrations) further quantities of one of the reactants.

The preceding equations have all assumed equal initial concentrations and linear reaction. If however, there is an excess of one of the reactants, the degree of polymerisation is reduced, hence the overall molecular weight is also reduced.

Impurities can significantly reduce molecular weight as well. However, for most surface coating applications only relatively low molecular weight condensation polymers are used and this is not a serious problem. It is however, a major problem in the fibre industry, where molecular weights greater than 100,000 are essential for fibre forming properties.

Using the Flory approach, consider the condensation reaction between functional groups A and B, either on the same molecule denoted $A-B$, or on separate bi-functional molecules denoted $A-A$ and $B-B$.

Either:

$$nA-B \ \rightarrow \ A-BA-BA-BA-B$$

or

$$A-A \ + \ B-B \ \rightarrow \ A-AB-BA-AB-B$$

If the number of A groups initially present is denoted as N_A and B groups as N_B, then let $r = N_A/N_B$ and p=fraction of A groups which have reacted at time t. The total number of units is $\dfrac{(N_A+N_B)}{2} = \dfrac{N_A\,(1+1/r)}{2}$

The number of chain ends is:

$$2\,N_A\,(1-p) + (N_B - N_A) \;=\; N_A\,[2\,(1-p) + (1-r)/r]$$

If there are only linear molecules present, then the number of molecules is half the number of chain ends, thus:

$$\bar{x}_n \;=\; \frac{1+r}{2r\,(1-p) + 1-r} \;=\; \frac{\text{number of units}}{\text{number of molecules}}$$

For equal numbers of functional groups A and B, r=1 and the equation reduces to the following:

$$\bar{x}_n \;=\; \frac{1}{(1-p)}$$

For complete reaction (i.e. p=1):

$$\bar{x}_n \;=\; (1+r)\,/\,(1-r)$$

These equations indicate the significance of equivalent concentrations of functional groups.

The further r is from unity the lower the molecular weight, irrespective of whether it is an acid or hydroxyl excess. In addition the degree of polymerisation is affected.

Flory showed that any deviation of r from unity introduces an error in $\bar{x}_n$ which increases, with increasing molecular weight, as the square of $\bar{x}_n$. At $\bar{x}_n$ of 200 an unaccounted loss of 0.1% of either A−A or B−B introduces a 10% error in the calculated molecular weight. This has very important consequences in typical condensation surface coating polymers, with their associated losses of volatile reactants (e.g. glycols) during processing. These are considered in later chapters.

The preceding equations can be applied to a mixture of bi-functional molecules all of the same class of functional groups (e.g. a mixture of ethylene glycol and neopentyl glycol). The mean molecular unit then depends upon the ratios of the different reactants.

The equations also hold if monofunctional reactants are present, provided the expression for r is modified to:

$$\frac{N_A}{(N_B + 2N_{(mono)})}$$

where $N_{(mono)}$ is the number of monofunctional reactants.

End group analysis (e.g. acid value) can be used for determining molecular weight. The growth of the polymer chain goes through odd and even stages with varying end groups, which can be illustrated as follows:

$$A-A + B-B \rightarrow A-AB-B \rightarrow A-(AB)_m-B$$

or

$$A-(AB)_m-B + A-A \rightarrow A-(AB)_{\overline{(m+1)}}A$$

$$A-(AB)_m-B + B-B \rightarrow B-(BA)_{\overline{(m+1)}}B$$

The type of molecules present at equilibrium depend upon the component in excess, and for equal initial concentrations there are equal types of molecules. Obviously there comes a point where oligomers can and do react with each other.

$$\bar{x} \text{ mer} + \bar{y} \text{ mer} = (\bar{x}+\bar{y}) \text{ mer}$$

The difference between $\bar{x}_n$ and $\bar{x}_w$ can be summarised as:

$$\bar{x}_n = \frac{1}{1-p} \; ; \; \bar{x}_w = \frac{1+p}{1-p}$$

where $M_n = M_o\bar{x}_n$

The discussion so far has been limited to linear polymers, but for most surface coating applications some degree of branching which stops short of cross-linking (as in an alkyd), is necessary to reduce the amount of polymer cross-linking necessary for film formation. This increases the film forming efficiency of the system for any given time and temperature. Most condensation polymers used in surface coating applications contain tri- or even tetra functional reactants, which have been added to introduce branching into the polymer chains.

The extent of reaction during processing has to be carefully controlled to avoid gelation in the reactor, whilst obtaining the maximum required degree of branching and molecular weight.

The preceding discussions have been restricted to linear polymerisation without any side reactions occurring. Whilst different functional groups are present on either end of the molecules, it is possible for either inter or intra molecular reaction to occur. This can result in ring formation and the length of the chain between the functional groups affects the probability of ring formation, but this is normally only small. There are other side reactions also occurring which do not take part in polymer formation.

There are numerous approaches to predicting the onset of gelation. They vary in degrees of sophistication but all are only approximations. The simple approach of Flory will be outlined and then compared with a more complex equation. The application of these predictive theories is illustrated in the Alkyd chapter.

Consider reacting a tri-functional monomer with the two bi-functional ones.

$$A-A + A-\begin{matrix} A \\ \\ A \end{matrix} + B-B$$

$$\rightarrow \quad -A\ [B-BA-A]_i\ B-BA-$$

Where **i** can be from 0 to ∞ depending upon the amount of tri-functional monomer present.

The principle of equal reactivity need not and does not necessarily apply to tri and higher functionality monomers. The secondary alcohol of glycerol is significantly less reactive than the primary ones. The four functional groups in penta erythritol are all of equal reactivity.

In theory at least a strictly bi-functional system cannot gel (i.e. form three-dimensional networks), but it can form infinite length linear chains.

Assume equal reactivity of all functional groups.

Let p_A represent the fraction of all A groups which have reacted and similarly for p_B. Let p represent the ratios of A's (reacted and unreacted) belonging to branch units to the total number of A groups present. The probability that B has reacted with a branch unit is $P_B p$. The probability that it is connected to a bi-functional A$-$A unit is $P_B (1-p)$. If the above reaction scheme is considered, the probability that the A group of a branch unit is connected to the shown sequence of units is:

$$P_A\ [P_B(1-p)P_A]\ ^iP_Bp$$

Consider now the probability that the chain ends in a branch unit (denoted α) and sum over all values of i to include all molecules. Then:

$$\alpha = P_A P_B p \ / \ [1 - P_A P_B (1-p)]$$

Let the ratio of A to B groups be represented by r where $r = P_A/P_B$ then:

$$\alpha = rP_A^2 \ p \ / \ [1 - rP_A^2 (1-p)]$$

or

$$\alpha = P_B^2 \ p \ / \ [r - P_B^2 (1-p)]$$

Under normal conditions r and p are determined by the proportions of the initial reactants. The unreacted A or B (or both) are determined during processing. The special cases outlined by Flory are as follows:

When there are no $A-A$ units, thus $p = 1$

$$\alpha = rP_A^2 = P_B^2/r$$

When A and B groups are in equivalent amounts $r = 1$, and $P_A = P_B = P$, thus

$$\alpha = P^2 \ p \ / \ [1 - P^2 (1-p)]$$

Consider a bi-functional and f-functional units $R - A_f$ where A can condense with A

$$\alpha = Pp \ / \ [1 - p(1-p)]$$

These equations are also applicable to tetra functional or higher functional systems. However, they are not directly applicable to molecules containing functional groups of different reactivity (e.g. glycerol).

As a general rule α can be calculated from the type and ratio of reactants and extent of reaction. There is a critical value of α denoted α_c at which the formation of an infinite network is possible. The critical value has been deduced as follows by Flory.

Consider a tri-functional branching unit and equal reactivity. Each chain which terminates in a branch is succeeded by two more chains. If both of these terminate in branches, four more chains are formed and so on. If $\alpha < \frac{1}{2}$ there is less than an even chance that each chain will lead to a branch unit and there is a greater than even chance that it will end at an unreacted functional group. The network cannot continue indefinitely and is therefore limited, and chain termination exceeds chain branching.

However, when α is greater than $\frac{1}{2}$, there is a better than even chance of each chain producing two chains, and these two chains will produce 4α new chains and n chains produces $2_n \alpha$ new chains. Therefore, branching of successive chains continues the structure indefinitely. Thus, for tri-functional systems a value of $\alpha = \frac{1}{2}$ represents the critical value of α for gelation (i.e. $\alpha_c = \frac{1}{2}$).

Even at the onset of gelation there are many separate molecules still present and this condition persists between $\frac{1}{2} < \alpha < 1$. The non-network material is termed a sol. The sol consists of monomer units, oligomers and polymers with functional groups which are unable to react further (i.e. dead ends). The number average degree of polymerisation is as follows:

$$\bar{x}_n = \frac{\text{number of units}}{\text{number of molecules}} = \frac{f(1-p+1/r)+2c}{f(1-p+1/r-2P_A)+2c}$$

Thus, $\bar{x}_n$ is not large at the gel point nor is it increasing rapidly. It cannot be over stressed that for a normal condensation reaction, onset of gelation does not mean that there is only one large infinite network present. It is gradual, and growth of the infinite network continues after the gel point has been reached. There is a mixture of small, large and infinite network molecules present.

Consider now the general case of infinite network formation which becomes possible, when the expected number of chains which will succeed n chains through branching by some of them exceeds n. Let f represent the functionality of the branching unit. Gelation occurs when $\alpha(f-1) > 1$. Therefore:

$$\alpha_c = 1/(f-1)$$

When there is more than one type of branching unit, the value of f is an average value obtained from the number of functional groups of each and their relative molar concentrations. The simplest theory to predict if gelation is likely to occur just calculates the functionality of the system.

If it is equal or greater than two then gelation may occur. Indeed, if it approaches two (e.g. 1.98) gelation is possible. It should be noted that this approach can only be used as a guide.

Functionality for the overall system (denoted F) is defined as the number of functional groups which are capable of entering into reaction per molecule present. Two examples will illustrate this:

$$m\,A-A + n\,B-B + q\,A\overset{A}{\underset{}{\perp}}A \;\rightarrow\; \text{polymer}$$

i) With exact equivalence of functional groups

$$F = \frac{2m+2n+3q}{m+n+q}$$

which is >2, therefore gelation will probably occur at a given degree of reaction.

ii) With an excess of A groups such that $2m+3q>2n$. For convenience let $m=q=n$,

then

$$F = \frac{2m+2n}{m+q+n} = \frac{4n}{3n} = 1.33$$

Only 2 n's worth of functional A groups are capable of reaction. The excess is neglected in calculating the number of functional groups.

Note that the same basic reaction was used for both examples. The only thing that varied was the relative initial concentrations of reactants. Obviously, if loss of a reactant occurs during processing, then this critical condition can be attained even though there was sufficient excess initially. If only a small excess if used, e.g. $(n+1)$ then as n increases, F increases until at $n \rightarrow \infty$, $F \rightarrow 2$. The onset of gelation can also be considered against the degree of polymerisation.

Carothers derived an equation for the number average degree of polymerisation n, which is obtained at various extents of reaction p, when the average number of functional groups per monomer molecule is f.

$$p = \frac{2}{f} - \frac{2}{n\,f}$$

At gelation n will be very large, so the extent of reaction at gelation p_{gel} is given by:

$$P_{gel} = \frac{2}{f}$$

It is therefore possible to calculate the theoretical extent of reaction at which gelation will occur by calculating f, the functionality of the system.

In practice the value of α_c is normally different to $1/(f-1)$ and there are many theories claiming to represent what really happens. α can be considered as the branching coefficient which is the probability of passing along one sequence of bonds from a randomly chosen branch unit to another branch unit (given that it has already reacted). Given that one group has reacted there are $(f-1)$ possible routes to the next unit, thus α $(f-1)$ is the probability that the next unit is reached, and when this is unity $\alpha=\alpha_c$ and gelation can commence.

For non-linear condensation measured values of α_c are always greater than those calculated from $1/(f-1)$ due to intra molecular reactions, and this deviation is normally significant. Some intra molecular reactions form rings, others 'backbite' (this can strictly be considered a ring). Stepto et al (Colloid & Polymer, Sec. 258, pps. 663−674 (1980) and p. 69 (1974)) have critically reviewed some of the theories concerning gelation and their significance.

Theories differ in their prediction of the gel point and for many surface coating resins the formulator is trying to prepare resins as close to the gel point as is possible, without losing the desired properties. Thus, knowledge of the gel point of any formulation is vital for two reasons, which are i) to stop gelation in the reactor, and ii) to maximise the performance of the resin.

Three theories for predicting gelation are as follows:

   i) Flory $\alpha_c = 1/(f-1)$

   ii) Kilb $\alpha_c = 1/[(f-1)(1-\lambda_k)]$

   iii) Frisch $\alpha_c = 1/[(f-1)(1-(1-\lambda_c)\lambda_F)]$

Where λ_k and λ_F are Kilb and Frisch constants which are ring forming parameters. The approximations differ in that Frisch assumed symmetrical growth from a central branch unit, whilst Kilb used a linear sequence of branch units. When λ_F and λ_k are both small, Stepto showed that their molecular interpretation was equivalent and this is the probability that a sequence of bonds from a chosen branch unit leads not to a new branch unit, but to one already connected to the chosen unit. λ is the total probability of intra molecular reaction for a branch unit and can be written as a sum of probabilities over all sizes of ring structure.

Relationships can then be derived for relating the concentration of internal and external reactive groups to λ. This then involves chain statistics and the reader is recommended to consult the relevant papers of Stepto, Flory and Stockmayer.

Another theory which could be used is that of Stockmayer (I. Pol Sci IX, (1) pps. 69−71 (1952)), which in the general form is:

$$(P_A P_B) \text{ gel} = \frac{1}{(f_A - 1)} \quad \frac{1}{(g_B - 1)} = \alpha_c$$

Where p_A is the fraction of A groups which have reacted and similarly for p_B. Effective rather than stoichiometric functionalities are assigned to the reactants in an effort to overcome the intra molecular reactions, and it is assumed that the principle of equal reactivity exists, then:

$$f_A = \frac{\Sigma_i f^2_i A_i}{\Sigma_i f_i A_i}$$

$$g_B = \frac{\Sigma_j g^2_j B_j}{\Sigma_j g_j B_j}$$

Where f_i represents the functionality of reactant A_i which is at molar concentration A_i and similar for g_j. The typical application of prediction of gel points is illustrated in the Alkyd chapter.

Chapter II

Glyceride Oils

Chapter II

Glyceride Oils

Glyceride oils are naturally occurring vegetable and marine animal oils. They occur as tri-glycerides of long chain fatty acids and are the reaction products of one molecule of glycerol and three molecules of a fatty acid. The chain length of the acid can vary from C_{12} to C_{22}, but C_{18} is the commonest species.

THE STRUCTURE OF GLYCERIDE OILS

Glyceride oils have the general structure:

$$
\begin{array}{l}
CH_2 - O - C - R^1 \\
\qquad\qquad \| \\
\qquad\qquad O \\[4pt]
CH\ \ - O - C - R^2 \qquad \text{Where } R^1, R^2 \text{ and } R^3 \text{ are the fatty acid residues}\\
\qquad\qquad \| \\
\qquad\qquad O \\[4pt]
CH_2 - O - C - R^3 \\
\qquad\qquad \| \\
\qquad\qquad O
\end{array}
$$

Although having the same general formula, the fatty acids can vary considerably in terms of:

    i) The number of $C = C$ double bonds (degree of unsaturation)

    ii) The relative positions of the double bonds (degree of conjugation)

    iii) The presence (or absence) of polar groups (such as OH or $C = O$) on the carbon backbone.

The variations in properties, encountered with the different oils, are a function of the variation in fatty acid structure. Since more than one type of fatty acid can be present per oil molecule, the properties of a particular oil can be directly related to the fatty acid composition. Some of the more important fatty acids are listed below and their compositions are compared in the table on page 52.

Lauric Acid

$$CH_3 - (CH_2)_{10} - COOH$$

Myristic Acid

$$CH_3 - (CH_2)_{12} - COOH$$

Palmitic Acid

$$CH_3 - (CH_2)_{14} - COOH$$

Stearic Acid

$$CH_3 - (CH_2)_{16} - COOH$$

Oleic Acid

$$CH_3 - (CH_2)_7 - CH = CH - (CH_2)_7 - COOH$$

Ricinoleic Acid

$$CH_3 - (CH_2)_5 - \underset{\underset{OH}{|}}{CH} - CH_2 - CH = CH - (CH_2)_7 - COOH$$

Linoleic Acid

$$CH_3 - (CH_2)_4 - CH = CH - CH_2 - CH = CH - (CH_2)_7 - COOH$$

Linolenic Acid

$$CH_3 - CH_2 - CH = CH - CH_2 - CH = CH - CH_2 - CH = CH - (CH_2)_7 - COH$$

Elaeostearic Acid

$$CH_3 - (CH_2)_3 - CH = CH - CH = CH - CH = CH - (CH_2)_7 - COOH$$

Licanic Acid

$$CH_3 - (CH_2)_7 - CH = CH - CH = CH - CH = CH - (CH_2)_4 - \underset{\underset{O}{\|}}{C} - (CH_2)_2 - COOH$$

THE CLASSIFICATION OF GLYCERIDE OILS

Glyceride oils are usually classified as drying, semi-drying or non-drying, and these divisions reflect the ability of the oil to 'air-dry', i.e. to form a coherent film on exposure to the atmosphere. This ability is directly related to the fatty acid composition.

In general, when fatty acids containing at least two double bonds are present, the oil will react with oxygen from the air to form a cross-linked network, i.e. the oil is said to 'air-dry'. The speed with which this occurs is greater if the double bonds are conjugated.

Drying oils will eventually form a 'tack free' film, whereas semi-drying oils form films that are never completely 'tack free'. Non-drying oils are unable to form a cross linked structure by air oxidation.

J.H. Greaves attempted to quantify the air drying ability of an oil by using a drying index. This drying index is defined as the percentage of linoleic acid present, plus two times the percentage of linolenic acid present.

So that, put in another way:

Drying Index = (% linoleic acid + 2 × % linolenic acid)

A drying index of greater than 70 indicates a drying oil.

Iodine value determinations can be used to measure the degree of unsaturation of an oil and this is useful in predicting the drying nature of the oil. An Iodine value of about 160 would normally indicate a drying oil.

The Iodine value of conjugated oils are best determined by the Woburn Method, since the more commonly used Wijs Method is quantitative only for non-conjugated systems.

THE CHEMISTRY OF THE DRYING PROCESS

In general, conjugated and non-conjugated systems 'dry', or film form, by cross-linking. The simplest approach is to postulate oxygen attack at the site of the activated methylene, which is alpha to the $C = C$ bond. This gives rise to hydroperoxide formation which subsequently breaks down with the formation of a link to a neighbouring fatty acid chain. The process differs somewhat for non-conjugated and conjugated systems as illustrated below:

Non-conjugated Systems

R〜CH = CH — CH₂ — CH = CH〜R

$\downarrow$ O₂

R〜CH = CH — CH — CH = CH〜R
                    |
                    O
                    |
                    O
                    |
                    H

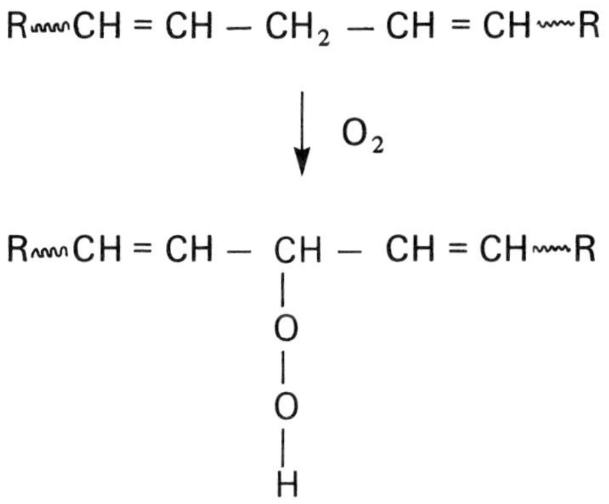

The hydroperoxides then decompose, by dissociation of the O—O bond, leading to a variety of reaction products including inter molecular linkage, i.e. cross-linking:

FATTY ACID COMPOSITIONS OF SOME OF THE MORE COMMON VEGETABLE OILS
(Approximate % Composition)

Fatty Acids	Unsaturation	Coconut Oil	Castor Oil	Grape Seed Oil	Linseed Oil	Oiticica Oil	Palm Oil	Palm Kernal Oil	Safflower Oil	Sunflower Oil	Soya Bean Oil	Tung Oil	Tall Oil*	Dehydrated Castor Oil*
8 Caprylic	$C_8H_{16}O_2$	6						3						
10 Capric	$C_{10}H_{20}O_2$	6						4						
12 Lauric	$C_{12}H_{22}O_2$	44				Contains 4% hydroxy acids		51						
14 Myristic	$C_{14}H_{28}O_2$	18					1	17						
16 Palmitic	$C_{16}H_{32}O_2$	11	2	9	6	7	48	8	8	11	11	4	5	
18 Stearic	$C_{18}H_{36}O_2$	6	1	4	4	5	4	2	3	6	4	1	2	
Oleic	$C_{18}H_{34}O_2$ (−2H)	7	7	20	22	6	38	13	13	29	25	8	48	9
Ricinoleic	$C_{18}H_{34}O_2$ (−2H)		87											
Linoleic	$C_{18}H_{34}O_2$ (−4H)	2	3	67	16		9	2	75	52	51	4	45	8
Linolenic	$C_{18}H_{30}O_2$ (−6H)				52				1	2	9	3		
Eleostearic	$C_{18}H_{30}O_2$ (−6H)											80		83
Licanic	$C_{18}H_{28}O_2$ (−6H)					78								
Approximate analytic constants														
Iodine Value		7.5−10.5	81−91	130−140	155−205	140−160	44−54	14−23	140−150	125−136	120−141	160−175	130−138	145−155
Saponification Value		250−264	176−187	185−195	188−196	186−193	195−205	245−255	188−194	188−194	185−195	189−195	192−194	200−204
Melting Point °C		23−26					27−50	24−26						
Titre °C		20−24		−17	19−21	42−47	40−47	20−28	15−18	16−20	20−21	36−37		
Uses		Short Oil non-drying Alkyd resins Epoxy resins	Plasticiser in ink systems Plasticising alkyds	Alkyd resins	Oleo-resinous varnish Long oil alkyd	Oleoresin varnish Alkyd resins	Short oil non-drying Alkyd resins		Alkyd resins	Alkyd resins	Alkyd resins	Oleo-resinous varnish Alkyd resins	Alkyd resins	Alkyd resins

*Typical tall oil fatty acids (2% rosin) and dehydrated castor oil shown for completeness

$$R - OOH \longrightarrow RO^{\bullet} + {^{\bullet}OH}$$

$$2\ R - OOH \longrightarrow RO^{\bullet} + ROO^{\bullet} + H_2O$$

Then:

$RO^{\bullet} + R^1H \longrightarrow ROH + {^{\bullet}R^1}$ radical transfer to another molecular

$R^{\bullet} + R^{\bullet} \longrightarrow R - R$ cross link

$RO^{\bullet} + R^{\bullet} \longrightarrow R - O - R$ cross link

$RO^{\bullet} + RO^{\bullet} \longrightarrow R-O-O-R$ cross link

Conjugated Systems

With conjugated systems, the hydroperoxide formed is predominantly, 1,4:

$$\sim\!\!\sim\!\!\sim CH = CH - CH = CH \sim\!\!\sim\!\!\sim$$

$$\downarrow O_2$$

$$\begin{array}{c} \sim\!\!\sim\!\!\sim CH - CH = CH - CH \sim\!\!\sim\!\!\sim \\ \quad\ |\qquad\qquad\qquad\quad\ | \\ \quad\ O\!\!-\!\!\!\rule{2.5cm}{0.4pt}\!\!\!-O \end{array}$$

The film formed differs from that formed by a non-conjugated structure in that the linkages are predominantly C—C suggesting a vinyl polymerisation mechanism:

$$\sim\sim\sim CH - CH = CH - CH \sim\sim\sim$$
$$\qquad\quad | \qquad\qquad\qquad |$$
$$\qquad\quad O\rule{2cm}{0.4pt}O$$

$$\qquad\quad OO^{\cdot}$$
$$\qquad\quad |$$
$$\sim\sim\sim CH - CH = CH - \dot{C}H \sim\sim\sim$$

$$\Big\downarrow \qquad + \sim\sim\sim CH = CH - CH = CH \sim\sim\sim$$

$$\qquad OO^{\cdot}$$
$$\qquad |$$
$$\sim\sim\sim CH - CH = CH - CH \sim\sim\sim$$
$$\qquad\qquad\qquad\qquad\qquad |$$
$$\sim\sim\sim CH - CH = CH - CH \sim\sim\sim$$
$$\qquad |$$
$$\sim\sim\sim CH - CH = CH - CH_2 \sim\sim\sim$$

In practice the length of the polymerisation chain is short because the probability of chain termination by oxygen is high.

A cross-linked oil film may be represented schematically as follows:

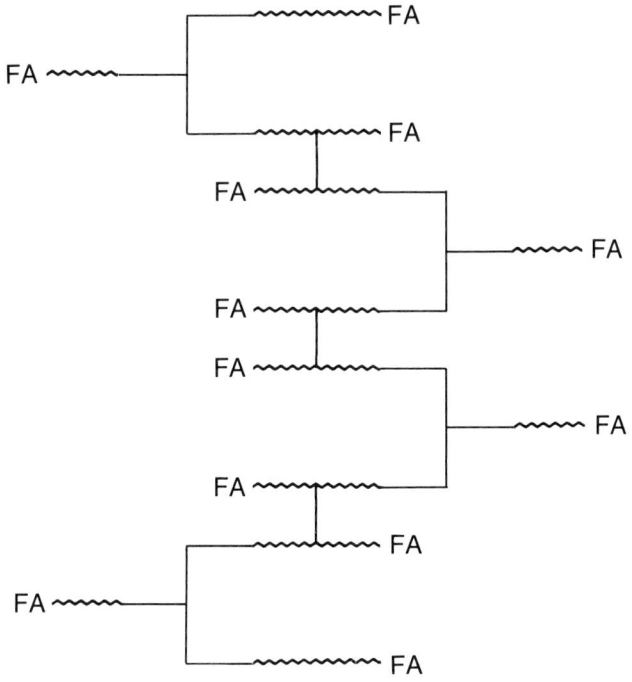

Where FA 〜〜〜 represents the fatty acid and ⌐ represents the glycerol portion of the oil molecule.

Fully saturated oils are susceptible to oxidation of the methylene group alpha to the ester group to form a hydroperoxide:

$$R^1 - CH_2 - \underset{\underset{O}{\|}}{C} - OR \xrightarrow{O_2} R^1 - \underset{\underset{OOH}{|}}{CH} - \underset{\underset{O}{\|}}{C} - OR$$

However, this reaction takes place much more slowly than the oxygen attack on the methylene group which is alpha to a carbon-carbon double bond. The methylene group alpha to the ester group therefore does not play such an important part in film formation, but it is a potential source of weakness which can eventually play a role in film degradation.

Factors affecting Film Formation

Several factors are of importance in influencing the cross-linking reactions.

The presence of certain transition metals increases the likelihood of peroxide breakdown by the formation of a redox system with the hydroperoxide:

$$ROOH + Co^{2+} \longrightarrow Co^{3+} + RO^{\bullet} + OH^-$$

$$ROOH + Co^{3+} \longrightarrow Co^{2+} + ROO^{\bullet} + H^+$$

They may also act as oxygen carriers to initiate free radical formation:

$$Co^{3+} + O_2 \longrightarrow Co^{3+} - O - O^{\bullet}$$

$$Co^{3+} - O - O^{\bullet} + R \sim\!\!\sim\!\!\sim CH_2 - CH = CH - R^1$$
$$\downarrow$$
$$Co^{3+} - OOH + R - \overset{\bullet}{C}H - CH = CH - R^1$$
$$\downarrow$$
$$Co^{3+} + R - \overset{\bullet}{C}H - CH = CH - R^1 + {}^{\bullet}OOH$$

These metals are collectively referred to as driers and are of great importance in surface coating systems. Without the use of these catalysts, film formation would be so slow as to be commercially unacceptable particularly with non-conjugated oils. There are three classes of driers:

i) **The Primary Driers.** These are organic salts (e.g. octoates, tallates and naphthenates) of transition metals, most usually Manganese or, Cobalt. They directly catalyse the oxidation of the oil, and are normally employed in amounts varying from 0.005 to 0.1% metal based on the oil. Their primary function is to promote rapid surface drying of the oil film.

ii) **The Secondary Driers or Promotors.** These are organic salts of zinc, calcium, lead, or barium. They do not have any direct catalytic effect on the oxidation when used on their own, but when used in conjunction with a primary drier, they act as synergists and considerably increase the rate of oxygen uptake of the oil.

iii) **Co-ordination Driers.** These are compounds of usually Zirconium or Aluminium which assist in the polymerisation process by the formation of co-ordination compounds. Zirconium acts as a through drier and is used as a replacement for the more toxic lead salts.

Promotors are normally used in amounts varying from 0.05 to 0.5% metal, based on the oil, and improve the stability of the drier/oil combination, reduce surface bloom and promote through drying of the film.

The Effect of Polarity

The polarity of the medium or solvent has an effect on the rate of hydroperoxide breakdown and hence the drying of the oil. In the presence of high boiling polar solvents, the hydroperoxides can be stabilised via hydrogen bond formation and hence drying is retarded:

$$R - OOH + R^1OH \rightleftharpoons R - OOH$$
$$\vdots$$
$$H - OR^1$$

This can occur also where polar groups are present in the oil either as natural impurities (e.g. phosphatides) or as part of the fatty acid structure (e.g. castor oil).

MODIFIED AND TREATED OILS

Raw glyceride oil, as extracted from its plant source, contains small quantities of impurities which can seriously effect its pigmentation and drying properties.

In particular phosphatides and carbohydrates are present, which will precipitate as a sludge when the oil is heated and lead to deleterious effects on colour, gloss and clarity of the surface coating.

It is normal therefore to refine the raw oil before use, and this is done by treating the oil with alkali or acid to precipitate the impurities. This technique gives rise to the acid and alkali refined grades of the glyceride oils.

The relatively slow drying speed of both raw and refined oils places limitations on the method of their incorporation into surface coating systems. It is common to upgrade the properties of the oil by direct chemical modification, or by blending or reacting with a synthetic resin. The latter techniques are described under the headings of the individual synthetic resins. Only direct chemical and structural modification of the oil will be considered in this chapter.

Polymerised Oils

One of the methods employed, to shorten drying time, is partly to polymerise or oxidise the oil prior to film application. The oil thus has an increased initial molecular weight and fewer cross-links are required to form a coherent film, i.e. the drying time is reduced.

Blown Oil

The oil is heated to about 130°C and air passed through it. Oxidation of the oil takes place under these conditions and the resulting polymerisation is controlled, by monitoring the accompanying increase in viscosity. The reaction may be catalysed by the addition of drier metals. The product so formed is termed a blown oil.

Bodied Oil or Stand Oil

Cross-linking or partial polymerisation can be achieved by heating the oil in the presence of a peroxide. In the case of highly conjugated oils (linseed and tung oil), the action of heat alone is sufficient to bring about polymerisation.

Bodied linseed oil is produced by heating the oil at about 260°C—300°C until the viscosity has increased to the desired value.

This technique results in a higher level of the preferred C—C bonded cross-links as opposed to the higher proportion of C—O—C bonded links obtained with blown oils.

Gas Proofing or Gas Checking of Tung Oil

These terms are applied to the thermal polymerisation of Tung oil. It is carried out to reduce the tendency of the films to 'frost' or 'gas check' due to the rapid surface drying phenomena associated with tung oil. The complete removal of surface wrinkling (particularly when the films are formed in poorly ventilated environments) is difficult, since tung oil tends to polymerise very rapidly and often gelation occurs before a completely 'gas-checked' system is obtained. To overcome the tendency of tung oil to gel during gas checking, the process is often carried out in the presence of a phenolic resin. This tends to slow the rate of polymerisation and give better control over the process.

Dehydrated Castor Oil

Castor oil, in its raw state, is a non-drying oil and is used in certain alkyd resins and ink formulations as a plasticiser.

Castor oil contains about 87% of Ricinoleic Acid (12—Hydroxy Oleic Acid) as the triglyceride and it is possible to remove the hydroxyl group from this acid, together with a hydrogen atom from a neighbouring carbon atom, to yield a conjugated fatty acid. In effect, this converts the non-drying castor oil into a drying oil.

The dehydration of castor oil is brought about by heating the oil at 265−270°C, in the presence of an acid catalyst, such as sulphuric acid or phthalic anhydride. Little reaction is obtained below 265°C, and a strong inert gas purge, or reduced pressure, is required to ensure the removal of the water of dehydration, and hence obtain a high level of conversion.

When the dehydrated castor oil is ultimately intended for use in alkyd resins, it is common to carry out the dehydration process in-situ with the alkyd condensation (see Chapter on Alkyd Resins, Chapter IV).

The dehydration process is monitored by the measurement of hydroxyl value, or more commonly by viscosity measurement. Since the viscosity of the hydroxy acid is higher than that of its dehydrated counterpart the viscosity of the oil decreases to a terminal minimum value during dehydration. The reaction is normally continued until at least 95% conversion of the ricinoleic acid has been obtained.

$$CH_3-(CH_2)_4-CH_2-\underset{\underset{OH}{|}}{CH}-CH_2-CH=CH-(CH_2)_7-COOH$$

$$\downarrow \quad \text{acid catalyst} \atop (270°C)$$

$$CH_3-(CH_2)_4-CH_2-CH=CH-CH=CH-(CH_2)_7-COOH$$
(Conjugated 9 : 11 isomer)

$+$

$$CH_3-(CH_2)_4-CH=CH-CH_2-CH=CH-(CH_2)_7-COOH$$
(Non-conjugated 9 : 12 isomer)

$+H_2O$

Two isomers of dehydrated ricinoleic acid are possible, the conjugated 9:11, and the non-conjugated 9:12 configurations. In practice only about 25−30% of the dehydrated oil is in the conjugated 9:11 form.

Despite this, the drying speed of dehydrated castor oil is still attractive for use in surface coating systems, since the drying speed lies somewhere between that of the semi-drying oils (soya, safflower etc.) and that of the drying oils.

Dehydrated castor oil dries to give a film with a very slight residual 'after-tack' and this is attributed to the formation of a small amount of estolide during dehydration.

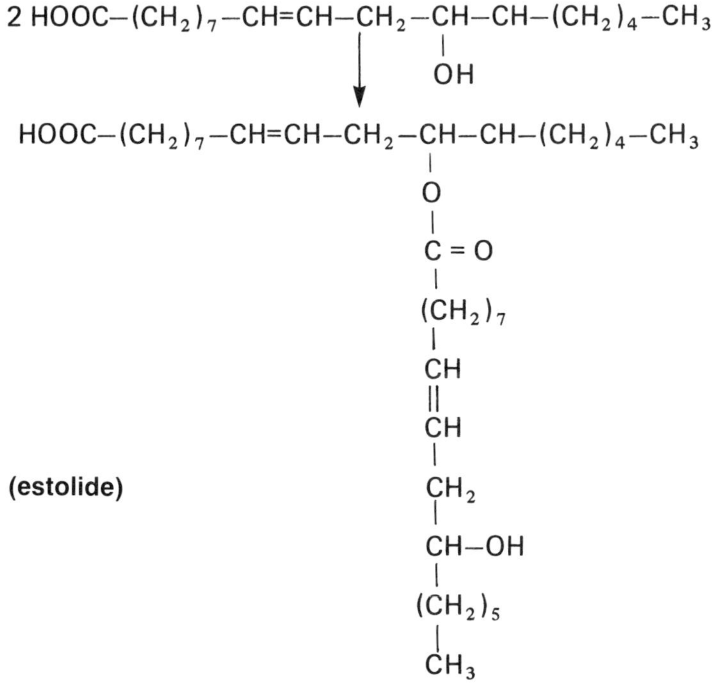

$$2 \; HOOC-(CH_2)_7-CH=CH-CH_2-\underset{\underset{OH}{|}}{CH}-CH-(CH_2)_4-CH_3$$

$$\downarrow$$

$$HOOC-(CH_2)_7-CH=CH-CH_2-CH-CH-(CH_2)_4-CH_3$$

(estolide)

$$\begin{array}{c}
| \\
O \\
| \\
C = O \\
| \\
(CH_2)_7 \\
| \\
CH \\
\| \\
CH \\
| \\
CH_2 \\
| \\
CH-OH \\
| \\
(CH_2)_5 \\
| \\
CH_3
\end{array}$$

Dimerised Fatty Acids

Dimerisation of unsaturated fatty acids can be carried out, using free radical initiators such as di-tertiary butyl peroxide.

Dimerisation is brought about by free radical attack on the methylene groups alpha to the $C = C$ double bond.

$$\sim\sim\sim CH = CH - CH_2 \sim\sim\sim$$

$$\downarrow \text{Peroxide}$$

$$\sim\sim\sim CH = CH - \overset{\bullet}{C}H \sim\sim\sim$$

$$\downarrow$$

$$\sim\sim\sim CH - CH = CH \sim\sim\sim$$
$$|$$
$$\sim\sim\sim CH = CH - CH \sim\sim\sim$$

The commercial dimerisation of unsaturated fatty acids, such as linoleic acid, is carried out using catalyst activated clay. The resulting dimer acids contain cyclic structures which are used in the manufacture of polyamides and as modifiers in alkyd resin manufacture (see relevant chapters for details).

$$
\begin{array}{c}
COOH \\
| \\
(CH_2)_7 \\
| \\
CH \\
\diagup \quad \diagdown \\
CH \qquad CH - (CH_2)_7 - COOH \\
\| \qquad\quad | \\
CH \qquad CH - CH = CH - (CH_2)_5 - CH_3 \\
\diagdown \quad \diagup \\
CH \\
| \\
(CH_2)_5 \\
| \\
CH_3
\end{array}
$$

Maleinised Oils

The maleinisation of oils and fatty acids, is of particular importance, as a first step in the manufacture of water dispersible and water soluble systems. It is also a useful method of introducing controlled chain branching into an alkyd resin.

The reaction between maleic anhydride and conjugated fatty acids proceeds readily by a Deils Alder type adduction at moderate temperatures of about 120°C.

$$\sim\!\!\sim\!\!\sim CH_2 - CH = CH - CH = CH - CH_2 \sim\!\!\sim\!\!\sim$$

$$+$$

$$\begin{array}{c}
CH = CH \\
| \qquad | \\
C \qquad C \\
O \quad O \quad O
\end{array}$$

$$\downarrow$$

$$\sim\!\!\sim\!\!\sim CH_2 - CH - CH = CH - CH - CH_2 \sim\!\!\sim\!\!\sim$$

$$\begin{array}{c}
CH \longrightarrow CH \\
| \qquad\qquad | \\
C \qquad\qquad C \\
O \quad\quad O \quad\quad O
\end{array}$$

Where non-conjugated fatty acids are involved, the reaction proceeds by attack on the active methylene groups and requires temperatures of about 200°C. This type of reaction is more commonly encountered in resin manufacture than the Deils Alder type above.

$$- CH_2 - CH = CH - CH_2 -$$

$$+$$

$$\begin{array}{c}
CH = CH \\
| \qquad | \\
C \qquad C \\
O \quad O \quad O
\end{array}$$

$$\downarrow$$

$$- CH_2 - CH - CH = CH -$$

$$\begin{array}{c}
CH - CH_2 \\
| \qquad | \\
C \qquad C \\
O \quad O \quad O
\end{array}$$

THE PROPERTIES AND
USES OF GLYCERIDE OILS

Linseed Oil

Linseed oil contains a high proportion of unsaturated linoleic and linolenic acids. Consequently good air-drying properties are conferred onto systems containing linseed oil. Linseed oil also gives excellent durability but poor acid and alkali resistance and only fair water resistance.

Alkali refined Linseed Oil

Alkali refined linseed oil gives excellent durability, good colour and gloss retention, good pigment wetting properties and air dries rapidly to give hard films. The dried films are susceptible to yellowing, however, and paints based on alkali refined linseed oil are used mainly for exterior coatings. It is extensively used in air drying alkyd resins and also in urethanated oils.

Blown Linseed Oil

Blown linseed oil has improved drying properties, and excellent flow and gloss. It also has a high water tolerance and is often added to water sensitive systems. However, it is renowned for pigment separation problems on extended storage.

Bodied Linseed Oil or Linseed Stand Oil

Bodied linseed oil or linseed stand oil is similar in properties to blown oil but has better drying, colour retention and durability. In addition, it has good levelling properties and has far less tendency towards pigment separation. It is used extensively on its own or combined to form oleoresinous varnishes in printing ink systems where its good pigment wetting and drying properties are desirable. The rheological properties that can be obtained from bodied linseed oil are particularly suited to lithographic varnishes and air-drying letterpress inks.

Tung Oil

80% of the fatty acid content of tung oil is the conjugated eleostearic acid. As a result tung oil exhibits rapid air-drying properties. Indeed the surface drying is so rapid that wrinkling often results. The tendency to wrinkle can be partly overcome by 'gas checking' (see Chapter — Gas Proofing or Gas Checking of Tung Oil).

Coatings employing tung oil have excellent chemical and water resistance properties. Because of its tendency to polymerise rapidly at elevated temperatures, it is rarely used on its own in alkyd resin systems, but is normally used as the minor component in blends with semi-drying oils.

Tung oil is extensively used with phenolic resins or rosin esters in oleoresinous varnishes for printing inks, but these inks are not used for printing food wrappings because of the high penetrating odour given off when the oil is drying.

Castor Oil and Dehydrated Castor Oil

Raw castor oil is a non-drying oil. It is used as the oil component in plasticising alkyds. However, dehydrated castor oil has excellent drying properties, and forms films with excellent colour, colour retention, water resistance and durability but which are soft and exhibit 'after tack'. It has good chemical resistance and excellent pigment, wetting properties.

Dehydrated castor oil is used in alkyd and epoxy ester resins, often in combination with soya bean oil.

It is frequently used in vinylated alkyd resins where the additional hardness conferred by the vinyl polymer portion offsets the softness of the normal dehydrated castor oil film.

Soya Bean Oil and Sunflower Oil

These oils are very similar in fatty acid composition and are often used interchangeably. Both have excellent colour retention properties and film flexibility. They are semi-drying oils and are mainly used in alkyd resins.

Safflower Oil

This contains more conjugated fatty acids than soya bean oil and hence has superior drying properties. It is a semi-drying oil and finds some use in alkyd resins where improved air-drying, relative to soya types, is required.

Oiticica Oil

This contains a high proportion of the conjugated fatty acid—Licanic Acid. The oil is noted for its quick drying and good gloss and adhesion. It gives films with excellent rub resistance, water resistance and alkali resistance. The films are more brittle and have a greater tendency to yellow than those of tung oil and so, oiticica oil is frequently mixed with soya bean oil to improve film flexibility. Oiticica oil is used in letterpress and lithographic inks that are required to dry on non-absorbent surfaces. Inks are frequently made from this oil when scratch proof prints are required for film and foil substrates. It tends to be used interchangeably with tung oil in ink formulations.

Chapter III

Hard Resins, Oleoresinous Media and Varnishes

Chapter III

Hard Resins, Oleoresinous Media and Varnishes

Many of the early varnishes were either drying oils in solvent, or natural resins in solvent, e.g. shellac lacquers.

The first type relied on chemical cross-linking induced by oxygen to form a coherent, tack free film. The second type relied for film formation only on the evaporation of the solvent.

As ink and coating technology improved, however, it was found that better properties could be obtained from both types, by chemically modifying the oil or the natural resin.

It was also found that incorporating resins into oils, could improve the properties of the dried film.

There are now available large ranges of modified natural resins and synthetic resins, modified oils and oleoresinous media and varnishes. The major types will be discussed in this chapter.

NATURAL RESINS

Natural resins have always played a role in inks and surface coatings, but their position has declined in recent times, as they have been gradually replaced by synthetic resins. Rosin is a notable exception and a separate section of this chapter is devoted to rosin and its derivatives.

Some of the more important natural resins are briefly described below.

COPALS

Congo copal is a fossil vegetable resin. Manila copal is a resin exudation from the pine trees of Borneo and the Far East. Two grades are available. A soft grade is obtained by tapping trees whilst the harder grade is obtained as a recent fossil resin. Kauri copal is a fossil resin from New Zealand. Congo copals contain Congo copalolic acid.

$$C_{36} H_{58} (CO_2 H)_2$$

and Congo copalolic acid

$$C_{21} H_{32} (OH) CO_2 H$$

with some α and β resenes.

Kauri copals contain the acids:

Kaurolic $C_{11}H_{19}CO_2H$
Kawenolic $C_{16}H_{33}CO_2H$
Kawinic $C_{19}H_{15}CO_2H$
Agathic $C_{11}H_{28}(CO_2H)_2$

AMBER

This is a fossil resin found in central Europe, no longer used in the manufacture of varnishes.

DAMAR

This is a tree exudation.

Unlike the copals, acids are a minor constituent of Damar, which contains over 60% of resenes such as:

$$C_{11}H_{17}O, C_{31}H_{52}O \text{ and } C_{22}H_{28}O.$$

The major acid constituents are:

damarolic $HOC_{54}H_{77}O_3(CO_2H)_2$

dagincoleic $C_{20}H_{42}(CO_2H)_2$

and dagingenoleic $HOC_{13}H_{26}O_3CO_2H$

SHELLAC

This is a secretion of the insect (Coccus Lacca) which feeds on the sap of acacia and mimosa trees.

Shellac consists of complex mixtures, but a major constituent is wax. Gluten is also present.

Until recently shellac was a very important natural resin but rising costs, scarcity and inferior performance (when compared with some synthetic resins) have all contributed to its decreasing usage. Even modified shellacs are inferior to many of the new ink and coating systems.

ZEIN

This is a protein obtained from maize or maize gluten.

BITUMINOUS RESINS

Large quantities of bitumen are used in paints for water-proofing. Gilsonite is still widely used in oleoresinous ink formulation, but only in very dark colours.

Gilsonite is a hard black inert solid which will dissolve in oils when heated.

HYDROCARBON RESINS

There are innumerable commercially available hydrocarbon resins, which can be incorporated into varnishes or media for printing inks and paints. As a general rule, most surface coating resin companies purchase rather than manufacture hydrocarbon resins. This is frequently because the resins are related to and immediately downstream from by-products of the petro-chemical industry and they generally require the use of specialist plant for their commercial preparation.

Many companies actively involved in cracking 'crude oil', offer a selection of hydrocarbon resins. Alternatively, a form of a 'hydrocarbon prepolymer', sometimes referred to as an active hydrocarbon, is commercially available. The latter may be further polymerised or co-polymerised by the surface coating manufacturer. Active hydrocarbon resins normally contain some degree of unsaturation.

Hydrocarbon resins differ in their chemical types and the major ones will be considered here.

Hydrocarbon resins derived from petrochemicals, can be based on a mixture of a wide range of constituents and incorporating both aliphatic and aromatic groups.

Amongst these Hercules (1,2) offers a wide range of hydrocarbon resins including hydrogenated ones and terpenes. Also offered by Hercules is a range of polystyrenes which can find limited uses.

Other suppliers include Allbright & Wilson (e.g. Resins), Mitsuit Petrochem Ind. (Petrosin) (3). ICI Petrochemicals Division (4).

Many of the commercially available hydrocarbon resins have properties which make them suitable for partial or total replacement of rosin or rosin derivatives to achieve a price reduction.

Esso have fairly recently marketed a range of reactive hydrocarbon resins, which can be used by the surface coating industry, for incorporation into ink or paint media. They are designated Escorez and contain cyclo pentene structures. An example is Escorez 800 which has bi-cycloheptene and cyclopentene unsaturation and can be depicted as:

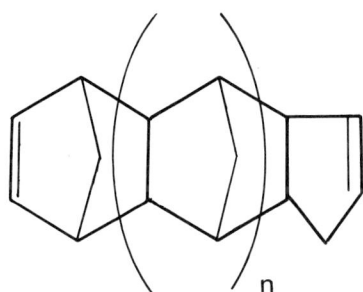

Both double bonds are capable of further reaction, including copolymerisation with drying oils, to form oleoresinous media.

Betaprene AC140 (Ex-Reichold) is a thermoplastic hydrocarbon resin which can be used in letterpress heat set, Web offset News and heat set, and roto-gravure inks.

WEB HEAT SET OFFSET BLACK INK

i)	Betaprene AC140	30
	Gilsonite	7
	Reprol (solvent)	27
ii)	Carbon black	20
	Talc 2060	2
	Solvent	8
iii)	Solvent	6

Dissolve i) in a ball mill, using heat if available; add ii) and mill to required grind; let down with iii).

COUMARONE AND INDENE RESINS

Coumarone and Indene occur in the coal tar naphtha fraction, distilling in the range 160−220°C.

Their structures are:

Courmarone   Indene H$_2$

The double bond in the five membered ring is available for polymerisation, either with itself, to form homopolymers or with other vinyl species to form copolymers. One of the most useful classes or resins of this series, is the resin formed by the copolymerisation of coumarone and indene.

Coumarone-indene resins are normally dark brittle solids or viscous liquids and find limited usage in inks. They are essentially neutral (i.e. acid value ca O mg KOH/g) which makes them ideal binders for metallic leafing or aluminium paints. A binder with an acid value can generate hydrogen with metallic pigments or corrode them.

With a low acid value they have poor pigment wetting, but offer good alkali and acid resistance and can be used in alkali resisting varnishes.

These resins can be used to form oleoresinous media, but care has to be taken to ensure that the temperature does not reach their polymerisation temperature of 300°C.

The resin darkens further on cooking so that sometimes a cold cut or milling operation is preferred.

Coumarone indene resins are incompatible with nitro cellulose but the presence of an ester gum substantially improves their compatibility with nitro cellulose.

Whilst they are soluble in both aliphatic and aromatic hydrocarbon solvents, they are insoluble in alcohols which obviously limits applications. It is necessary to use oxygenated solvents, like cellosolves, to achieve complete solution of these resins and in general they exhibit poor solvent release properties.

Among the commercial suppliers of coumarone indene resins are:

Britrez (5) — British Steel Corporation
VFT (6)   — Rutgerswerker AG Tar Chem Div

Coumarone indene resins are relatively inexpensive when compared with many of the resins used in the printing ink industry, so that sometimes, they are used in conjunction with other resins, to reduce the overall price.

Oleoresinous media based on these resins can be used for gravure inks.

Other resins in this category include polyindenes and polycoumarones although many suppliers tend to class all of these as coumarone indene resins.

Colour can be improved by hydrogenating the coumarone resins. The hydrogenation of coumarone resins leads to improved solubility, particularly in aliphatic solvents.

Styrene can be copolymerised with coumarone resin to form a useful range of resins and coumarone resin is often incorporated into phenolic based resins.

KETONE RESINS

Cyclic ketones or cyclohexanone resins tend to be brittle and of low colour.

The resins are prepared by an aldol type condensation of the ketone in the presence of an alkaline catalyst. The actual mechanism and structures have not yet been fully elucidated.

The majority of ketone resins are compatible with nitro cellulose. They also exhibit a good solubility in a wide range of solvents.

Some of the ketone resins and particularly the cyclic ones, can be incorporated into cold cut oleoresinous media.

Examples of using some of the Laporte cyclic ketone resins (MS2, MC2) are quoted here (7).

GRAVURE INK MEDIA

Nitro cellulose HX 30/50	10
MS2 (ex-Laporte)	4
Howflex SP (ex-Laporte)	6
Ethyl acetate	35
Butyl acetate	18
Sextone B (ex-Laporte)	15
Toluene	12
	100

GREEN CAR BODY LACQUER

Nitro cellulose AX 30/50	10.00
Howflex DBP (ex-Laporte)	4.00
Blown castor oil	2.00
Resin MS2 (ex-Laporte)	6.00
Yellow and blue pigments	4.57
Blanc fix	1.43
Methyl isobutyl ketone	7.00
Butyl acetate	7.50
Sextone B (ex-Laporte)	14.50
Toluene	25.00
Xylene	18.00
	100.00

WATER RESISTANT
CHLORINATED RUBBER PAINT

Chlorinated rubber	15
Resin MS2 (ex-Laporte)	6
Howflex SP (ex-Laporte)	9
Titanium di-oxide	15
Toluene	25
Sextone B (ex-Laporte)	10
White spirit	20
	100

Beckacite 1409 and 1410 (ex-Riechold) (8) are alcohol ketone aldehyde condensation resins which can be used in inks and varnishes.

ROTOGRAVURE 'E' OR
FLEXOGRAPHIC SOLVENT INK

* RBH 4902 black	(see below)	40
* * Nitrocellulose lacquer	(see below)	25
Beckacite 1409 (50% non-volatile content ethyl alcohol)		20
Ethyl alcohol		10
Polymekon wax compound		5
		100

* **Pigment for above**	
Pigment	25.0
S.S. N/C	12.5
Solvent	62.5
	100.0

** Nitrocellulose lacquer for above

¼ " S.S. N/C (70% dry)	43
Ethyl acetate	7
Ethyl alcohol	50
	100

ROTOGRAVURE OR
FLEXOGRAPHIC POLYAMIDE INK

* Polyamide varnish	(see below)	40
** Carbon black	(see below)	12
Isopropanol (91%)		8

Ball mill grind, check and add:

* Polyamide varnish	15
Beckacite 1409 (50% non-volatile content ethyl alcohol)	8
Polyethylene wax compound	5
Isopropanol	8
VM & P Naphtha	4
	100

* Polyamide varnish for above

Wolfamid 106	
Wolfamid 123	
Wolfamid 107	40
VM & P Naphtha	30
Isopropanol, 91%	30
	100

** **Pigment for above**

Mogul LV Cabot Corporation

Peerless 155 Cities Service

ROTOGRAVURE 'E' OR
FLEXOGRAPHIC SOLVENT INK

* RBH 4902 black	(see below)	40
** Nitrocellulose lacquer		25
Beckacite 1410 (50% non-volatile content ethyl alcohol)		20
Ethyl alcohol		10
Polymekon wax compound		5
		100

* **Pigment for above**

Pigment	25.0
S.S. N/C	12.5
Solvent	62.5
	100.0

Nitrocellulose lacquer for above

¼" S.S. N/C (70% dry)	43
Ethyl acetate	7
Ethyl alcohol	50
	100

ROSIN

Rosin occurs naturally as a resin or gum in pine wood. The exudation obtained by 'tapping' pine trees is known as 'gum rosin'. It is also obtained from the stumps of pine trees by shredding, solvent extraction and subsequent distillation of volatile materials, producing 'wood rosin' which is similar to gum rosin. Tall oils contain a large proportion of rosin, which is separated by distillation giving tall oil fatty acids and tall oil rosin.

As a natural product, there are likely to be large variations in the composition and properties of each batch of rosin. This is not only because of different sources but also because of variations in climate. Amongst the countries from which rosin can be obtained are USA, Canada, Portugal, China, Greece, Brazil, Spain and Scandinavia.

Rosin varies considerably in colour, from a pale yellow to almost black, and a wide range of colour is obtained from each country.

THE COMPOSITION OF ROSIN

Rosin consists of a mixture of various rosin acids, which are monocarboxylic acids of alkyl hydrophenanthrene, with a preponderance of abietic acid. In addition to the rosin content, there is a small proportion of fatty acids, unsaponifiable material and a very small quantity of sulphur containing compounds.

Some of the major rosin acids in rosin are as follows:

Abietic acid

Neo abietic acid

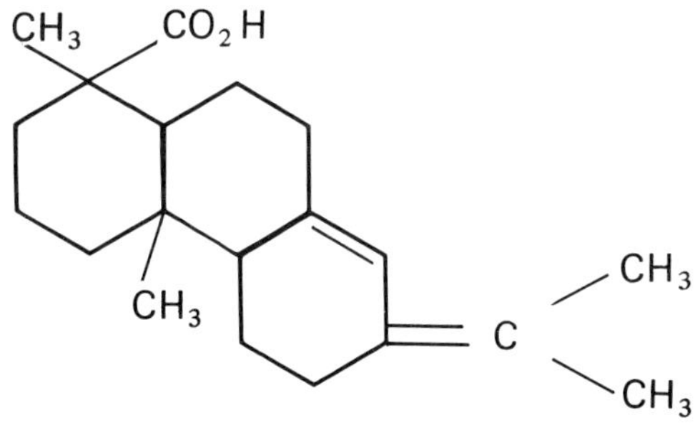

De hydro abietic acid

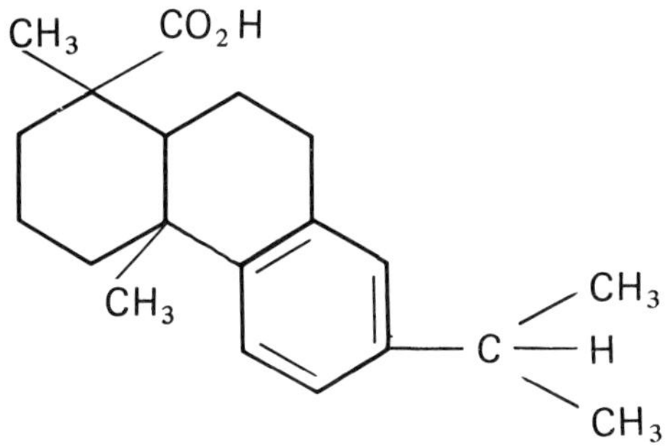

Palustric acid

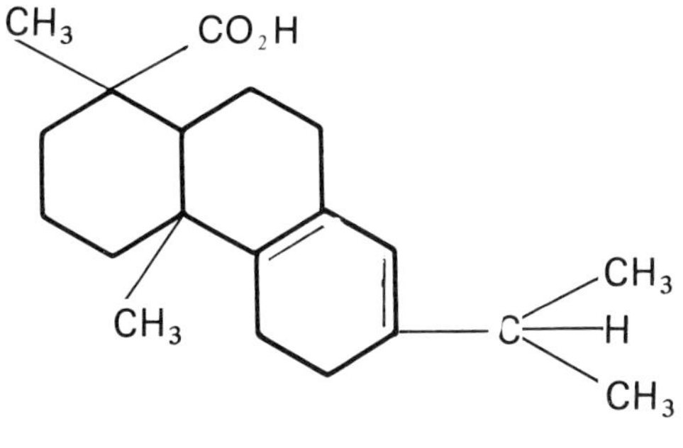

Dextro-pimaric acid

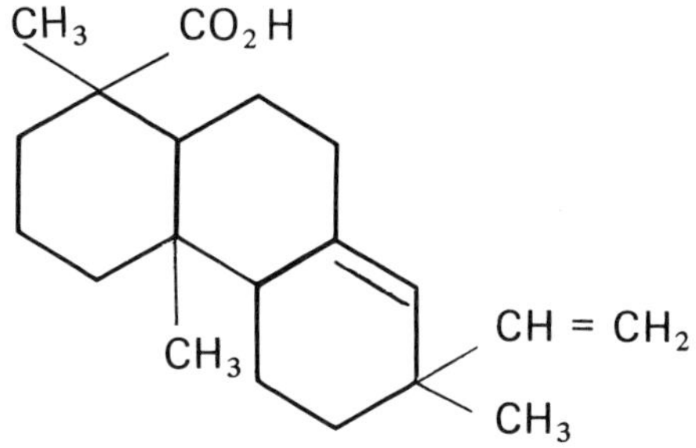

CH₃   CO₂H

CH = CH₂
CH₃
CH₃
CH₃

Iso-dextro-pimaric acid

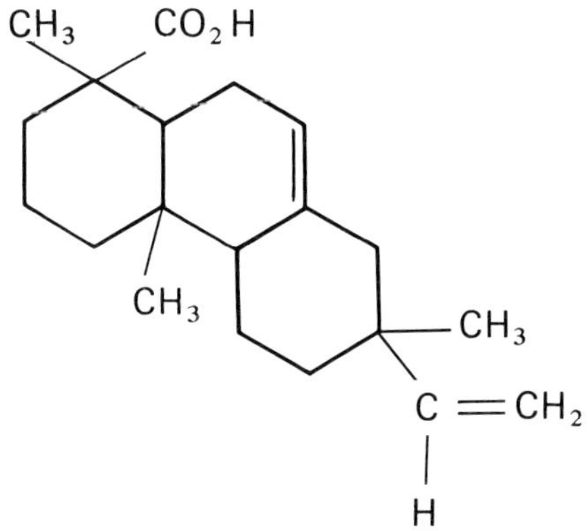

CH₃   CO₂H

CH₃
CH₃
C = CH₂
H

Heat can cause acids to convert between structures (e.g. abietic to leavo-pimaric).

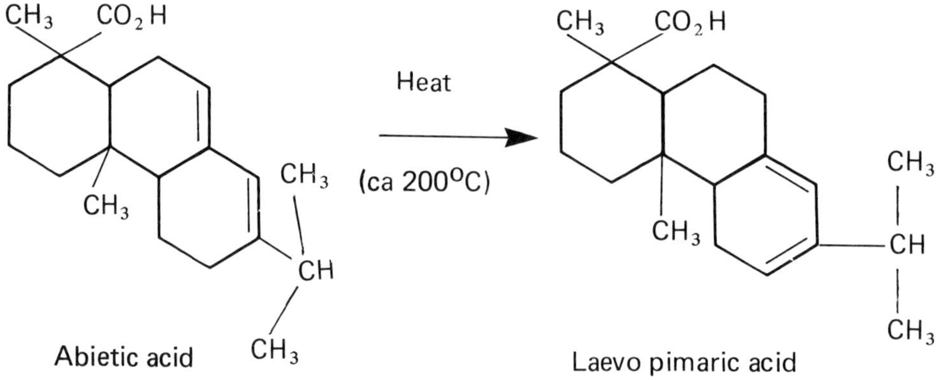

CH₃ CO₂H

Heat

CH₃
CH₃
CH₃
CH
CH₃

(ca 200°C)

Abietic acid

CH₃ CO₂H

CH₃
CH₃
CH
CH₃

Laevo pimaric acid

The approximate composition of the three types of rosin is:

Tall Oil Rosin	Laevo Pimaric Acid	Abietic, Neo-Abietic and Palustric Acids	Dehydro and Tetrahydro Abietic Acids	Dextro Pimaric Acids
Wood Rosin	<1	45–55	15–25	15
Gum Rosin	0–15	65–80	Traces	15–30
Tall Oil Rosin	<1	40–50	20–30	15–30

At room temperature, rosin normally consists of a substantial (ca 90%) amount of abietic acid, but there is a temperature dependent equilibrium concentration of the different acids.

The two major types of acid present in rosin are abietic types and pimaric types. The carboxyl group of both types is capable of undergoing similar reactions.

The two types differ in the position of the double bonds. The pimaric type acid does not have the conjugation of the abietic acid type. This, of course, means that the double bonds in pimaric and iso-pimaric acid are not as reactive as those in abietic acid, so that they cannot undergo Diels Alder reactions. However, they can still be maleinised (see the section on the Maleinisation of Rosin), and they are more resistant to oxidation and can be hydrogenated.

Unmodified rosins have a tendency to crystallise and oxidise. Crystallisation can be reduced by heating, polymerisation, partial neutralisation with sodium hydroxide, or reaction with dienophiles. Oxidation can be reduced by disproportionation (heating with catalysts to remove conjugated double bonds), hydrogenation, maleation and to some extent by polymerisation. Many of the above reactions are considered later in this chapter.

TYPICAL PROPERTIES OF ROSIN

Some of the typical properties of rosin can be summarised as follows:

Acid value 165–170 mg KOH/g

Saponification value 170–180 mg KOH/g

Melting point (B + R) 80°C.

ROSIN MODIFICATIONS

Rosin is a versatile component of many varnishes, imparting many properties desired in conventional ink and paint varnishes. The usefulness of rosin as a synthetic resin modifier derives from the following:

1. A carboxyl group which can be reacted to form esters, metal salts or converted into other groups.

2. Double bonds which can be hydrogenated, maleated or polymerised. Reaction with resole phenolics is believed to occur via the double bond. (See Chroman Ring formation in Chapter VI on Phenolic Resins.)

3. A molecular structure, containing aliphatic and alicyclic groups, which impart solubility in many of the organic solvents used in the surface coating industry. It has a solubility parameter (8) which lies between the values of many common solvents (e.g. white spirit and xylene).

The structure also imparts some degree of water insolubility and resistance, although where ester links are present, no great improvement in these properties is observed.

The reactions of rosin have been divided into four categories:

   i) Reaction of carboxylic acid

  ii) Reaction of double bond

 iii) Rosin modified phenolics

 iv) Miscellaneous rosin compounds.

THE REACTIONS OF ROSIN ACIDS (Carboxyl Group)

The carboxyl group of rosin acids is not as reactive as that in a fatty acid. This is because it is sterically hindered, being attached to a tertiary carbon atom, which is part of a cyclic ring attached to two other cyclic rings.

The carboxyl group undergoes most of the reactions that fatty acids undergo, except that frequently higher reaction temperatures are required, especially for esterification. This steric hindrance also imparts some degree of resistance to hydrolysis (both acidic and basic). The reactions of the carboxyl group which are of interest to the surface coatings chemist are:

  a) Esterification

  b) Soap formation (i.e. resinates)

  c) Decarboxylation.

Esterification

The carboxyl group reacts with mono and di-hydroxy alcohols, but the products have lower melting points than rosin. Poly hydroxy alcohols (i.e. polyols) like glycerol or pentaerythritol must be used to form resinous products with melting points above that of rosin.

Rosin has many useful properties, but suffers from being tacky. Reaction with glycerol, pentaerythritol, di-pentaerythritol or other polyols increases the melting point and improves the non-blocking characteristics. As a general rule reaction with di-pentaerythritol increases the melting point more than with pentaerythritol, which is significantly higher than with glycerol for the same equivalent amount of polyol.

Both glycerol and pentaerythritol rosin esters can be combined with phenolic resins or with maleic anhydride, or they can be used unmodified in oleoresinous media.

The simplest example of esterification is the formation of rosin ester gums:

ROSIN ESTER GUM
(Glycerol Esterified Rosin)

Rosin	89
Glycerol	11
	100

PROCESS

Charge the rosin and heat to 200°C under an inert gas atmosphere. Add the glycerol slowly and allow the temperature to rise to about 260°C. A 10% excess of glycerol above the stoichiometric amount is used to allow for process losses. The excess glycerol also increases the rate of reaction and negates the need to use esterification catalysts. Reaction is monitored by acid value measurement and by determination of the increase in melting point. Reaction is generally considered complete when the acid value is below 20 mg KOH/g. The resin is then chilled, discharged and allowed to solidify before being kibbled.

Esterification catalysts may be used to increase the reaction rate and shorten process times.

Catalysts include basic metal oxides or metals, acids, zinc or iron chlorides and calcium resinate.

A typical glycerol ester gum would have an acid value below 10 mg KOH/g and a melting point of 80−90°C. It is soluble in hydrocarbons, white spirit and drying oils, but is insoluble in alcohols.

Spirit soluble ester gums can be formed by stopping the reaction at a high enough acid value to ensure alcohol solubility.

The water resistance of ester gums is poor and their films tend to be soft. The higher acid value (e.g. 10−12 mg KOH/g) estergums are preferred in oleoresinous varnishes.

An alternative method of practical control is by continuing the reaction until a bead of the reaction mixture clears on cooling. This indicates the consumption of all the glycerol. Nonetheless acid and hydroxyl value controls together with melting point determination are to be preferred.

The reaction can be summarised below, where R represents abietic acid residue.

$$3\,RCO_2H \quad + \quad
\begin{array}{l} CH_2OH \\ | \\ CH\,OH \\ | \\ CH_2OH \end{array}
\quad \rightarrow \quad
\begin{array}{l} CH_2O_2CR \\ | \\ CHO_2CR + 3H_2O \rightarrow \\ | \\ CH_2O_2CR \end{array}$$

| Abietic acid | Glycerol | Glycerol ester gum |

Pentaerythritol Esterified Rosin

The reaction mechanism can be summarised as:

$$4R\,CO_2H + HOCH_2
\begin{array}{c} CH_2OH \\ | \\ -\,C - CH_2OH \\ | \\ CH_2OH \end{array}
\rightarrow RCO_2CH_2-
\begin{array}{c} CH_2O_2CR \\ | \\ C-CH_2O_2CR + 4\,H_2O \\ | \\ CH_2O_2,CR \end{array}$$

FORMULATION

Rosin	1 equivalent (CO_2H)
Pentaerythritol	1.02 equivalent (OH)

PROCESS

Rosin is heated and pentaerythritol is added at about 150°C. The reaction mixture is heated to 250°C–260°C. Process control is by the determination of acid value. A very small polyol excess of 5% maximum of pentaerythritol is used, otherwise the solubility of the ester in oil will be impaired. The above formulation requires that 98% (or better) purity pentaerythritol is used. This grade is commercially available in U.K.

The acid value is normally in the region 12–20 mg KOH/g and the gum has good oil compatibility but poor nitro cellulose compatibility. The melting points of pentaerythritol gums are correspondingly higher than glycerol ester gums. The films are harder and the gum has a higher decomposition temperature. Drying rate is increased due to lower solvent retention.

The molecular weight of the pentaerythritol rosin ester gum is 25% larger than that of a glycerol ester gum.

Rosin is essentially mono functional with respect to the acid group, therefore cross linked polymers cannot be formed from simple esters. It can also be used for chain stopping in alkyds.

Pentaerythritol, since it has four primary hydroxyl groups, reacts faster and more easily to completion than glycerol (two primary and one secondary hydroxyl group). The polyol losses with pentaerythritol are minimal under normal process conditions.

Thus for the majority of applications, where price is not the overriding factor, pentaerythritol is preferred to glycerol as the esterifying polyol.

Esterification Rosin Types

The dimers, trimer and polymers of rosin will all undergo esterification. The trifunctional maleic adduct can be esterified. Di-pentaerythritol can be used as a total or partial replacement (on an equivalent basis) for the polyols in any of the ester gums or modified rosin formulations where modified properties are required.

Soap Formation (Resinates)

Rosin acids being acidic, readily undergo reaction with bases forming metal resinates. Metals used, include sodium, potassium, zinc, calcium, lead, cobalt, manganese and aluminium.

Three different methods can be used for resinate formation:

  i) Precipitation (Double Decomposition) Process

  ii) Solvent Process

  iii) Fusion Process.

The Precipitation Process

A water soluble metal salt is added to an aqueous solution of the sodium salt of rosin. This precipitates the resinate, which is then filtered, washed and dried. If required, the resinate can be further purified by dissolving in solvent (either aromatic or aliphatic) and filtering etc.

The reaction can be summarised as follows:

$$2C_{19}H_{29}CO_2Na \quad + \quad 2(CH_3CO_2)Zn$$

Sodium resinate   Zinc acetate

$$\downarrow$$

$$(C_{19}C_{29}CO_2)_2 \; Zn \quad + \quad 2CH_3CO_2Na$$

Zinc resinate   Sodium acetate

Sodium resinate would be dissolved and filtered so that the non-reactive parts of rosin are removed. To this solution, an excess of zinc acetate would be added, to drive the reaction as close to completion as possible. Any excess would be removed during washing.

To form cobalt resinate, cobalt nitrate is used:

$$2C_{19}H_{29}CO_2Na \quad + \quad Co(NO_3)_2$$

Sodium resinate   Cobalt nitrate

$$\downarrow$$

$$(C_{19}H_{29}CO_2)_2Co \quad + \quad 2NaNO_3$$

Cobalt resinate   Sodium nitrate

Solvent Process

The rosin is dissolved in (normally) a hydrocarbon solvent and heated with a reactive metal compound. When the reaction is complete, solvent and other volatile materials are removed. Care must be taken to ensure complete reaction, otherwise there is a suspension of the metal compound. This will give rise to haziness. Reactive metal compounds include oxides, hydroxides, acetates and carbonates.

Reaction can be followed by monitoring acid value decrease and it should be noted that it may be necessary to filter insoluble reactive metal compounds from the reaction mixture, before determining acid value.

Fusion Process

The chemistry for the fusion process is similar to that of the solvent process. A reactive metal compound is added to molten rosin. The amount of rosin converted to resinate is less than when using the other methods. Over-reaction leads to the formation of an infusible gel. This can be overcome by using an aldehyde modified rosin, which allows complete conversion to the resinate:

LIME HARDENED ROSIN

Rosin	91.9
Castor oil	5.0
Slaked lime	7.0
Calcium acetate	0.1
	100.0

PROCESS

Lime hardened rosin, was one of the earliest forms of resinate. A paste of slaked lime and castor oil is prepared and added to molten rosin at 230°C.

A slow addition rate must be used to avoid frothing. When the addition is complete the temperature is raised to 250°C and calcium acetate is added.

When the acid value has decreased to the desired level the reactants are cooled and discharged prior to crushing.

Acid values below 40 mg KOH/g are seldom formed because of dispersion problems with the slaked lime. A combined zinc, calcium resinate will have a much lower acid value and higher melting point than when calcium alone is used.

The Uses of Resinates

Resinates raise the melting point of Rosin, whilst reducing the acid value. Typical examples would be:

	Melting Point (B&R)°C	Acid Value
Rosin	80	170 mg KOH/g
Calcium resinate	130	50 – 75 mg KOH/g
Zinc resinate	160	less than 10 mg KOH/g

The metal content would range from 2−10%. Mixed calcium and zinc resinates may contain 5% zinc and 2.5% calcium. Examples of this are resinates B and H from Warwick Chemicals. Leon Frenkel and Cornbrook also supply resinates.

Resinates B, H and Kelrez 42−455, with some of the Ennesin range are designed for use in gravure ink formulations with either aliphatic or aromatic solvents. Melting points are in the region of 160−185°C.

The potassium, sodium, ammonium and amine salts of rosin are used as emulsifiers. Limed rosin is soluble in mineral spirit and is used as a binder in printing inks and as a gloss oil in paints and varnishes. Zinc resinate is used in paints, varnishes and inks. It imparts gloss to varnishes, assists grinding and dispersion of pigment and produces better 'through dry' in films. Heavy metal soaps of rosin can function as driers for paints and varnishes.

The melting point of a resinate is roughly proportional to the amount of metal present. Resinates can be air blown to increase their melting point by about 20°C for the same metal content.

Decarboxylation

The carboxylic acids in abietic acid (i.e. rosin) will undergo decarboxylation at ca 300°C, forming a neutral rosin oil. Acid value and melting point is decreased. Decarboxylation can be promoted by using catalysts like zinc, chloride, sulphuric acid, phosphorous pentoxide and silicaceous earths at temperatures between 150−270°C. Rosin oil finds limited use in the ink and paint industries.

THE REACTION OF DOUBLE BONDS

Reactions which involve the double bonds of rosin acids include:

a) Isomerisation

b) Air oxidation

c) Disproportionation

d) Polymerisation

e) Maleinisation

f) Hydrogenation.

a) and b) Isomerisation and Air Oxidation have already been discussed.

Some of the reactions only occur with conjugated dienes and are limited to abietic type acid. Whilst the reaction of reactive phenolic resins (resoles) is believed to occur through the double bond, non rosin modified oil soluble phenolics will be considered separately.

Disproportionation

Disproportionation is used to improve the resistance of rosin to oxidation by air. Disproportionation is a dehydrogenation − hydrogenation reaction. The first step is dehydrogenation and the formation of an aromatic ring in the double bond abietic type of acid to form dehydroabietic acid.

The hydrogen which is removed in the first step then hydrogenates the two double bonds, forming di-hydro and tetra-hydro abietic acid.

Disproportionation will occur merely on heating but the use of catalysts like iodine, sulphur di-oxide or noble metals, improves the efficiency. However, at 300°C in the presence of a catalyst, care must be exercised, otherwise decarboxylation and complete dehydration can occur to give an aromatic hydro-carbon resin. The actual reaction mechanism is not as simple as would appear at first sight. Under the normal commercial conditions used for disproportionation, a higher than theoretical amount of dehydroabietic acid is produced. This means that additional dehydrogenation has occurred, perhaps indicating a loss of hydrogen through its possible escape as a gas.

The aromatic ring and hydrogenated structures are highly resistant to oxidation.

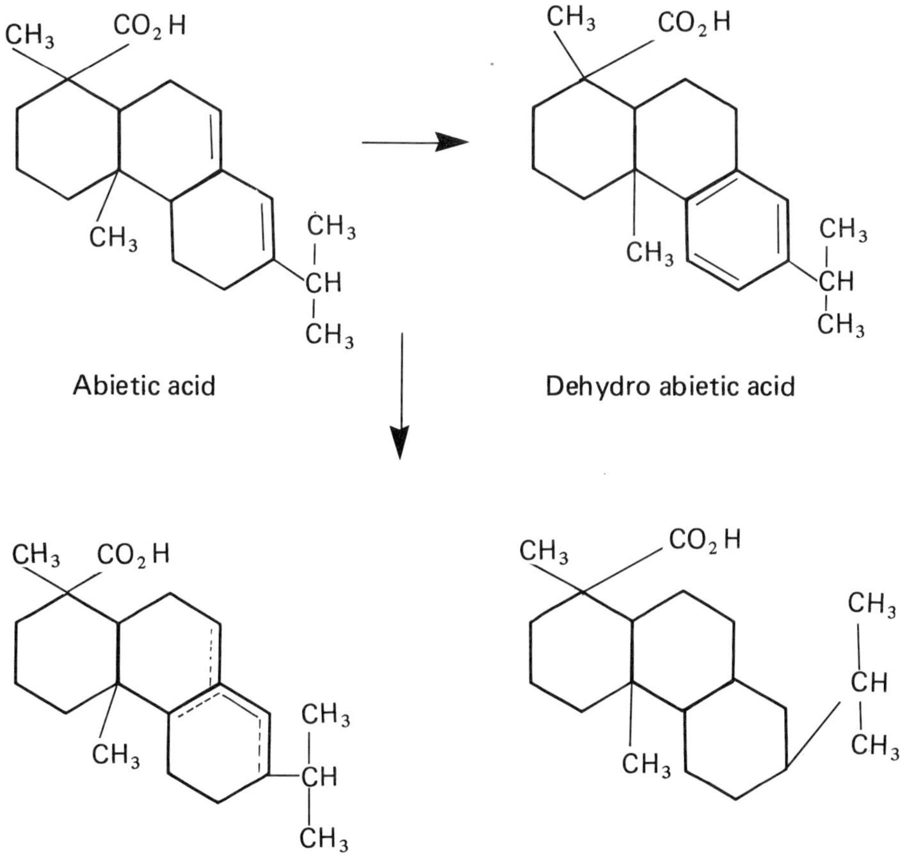

Abietic acid

Dehydro abietic acid

Di hydro abietic acid
(a mixture of isomers with
one double bond)

Tetra hydro abietic acid

Polymerisation

Rosin can be polymerised, forming dimers, trimer and polymers, using Lewis acid type catalysts (e.g. Zinc chloride, phosphoric or sulphuric acids).

The polymerised rosins available commercially are mixtures of unpolymerised rosin, dimerised rosin and sometimes trimerised rosin depending on the catalyst used. The mechanisms of polymerisation are still unclear but a Diels Alder type of reaction is believed to occur.

The polymerised rosin still retains its acidic reactivity and can undergo esterification and other reactions. Examining some of the properties of a range of rosin esters, available from Leon Frenkel, illustrates some of the trends discussed.

Rosin Type	Esterifying Polyol	Melting Point (B&R)°C	Acid Value mg KOH/g	White Spirit Tolerance
Rosin	Glycerol	84 – 87	6	1:10
Rosin	Glycerol	84 – 88	10	1:10
Rosin	Pentaerythritol	95 – 105	10	1:10
Dimerised rosin	Pentaerythritol	155 – 160	15	1:10
Hydrogenated rosin	Glycerol	73 – 78	10	1:10
Hydrogenated rosin	Pentaerythritol	87 – 92	15	1:10
Polymerised rosin	Glycerol	113 – 116	10	1:10
Polymerised rosin	Pentaerythritol	115 – 122	10	1:10
Disproportionated rosin	Glycerol	65 – 73	13	1:10
Disproportionated rosin	Pentaerythritol	80 – 90	15	1:10
Hydrogenated rosin	Pentaerythritol	96 – 100	15	1:10
Copal	Glycerol	95 – 100	7	1:10
Copal	Pentaerythritol	105 – 110	9	1:7

Using pentaerythritol increases the melting point, whether it is a rosin or modified rosin which is esterified. Replacing rosin by dimerised or polymerised rosin increases the melting point.

Hydrogenated rosin results in a lowering of the melting point, as do disproportionated rosins. The resins so produced however, have better colour stability. Replacing rosin by copal, results in an increased melting point.

Maleinisation of Rosin

Maleic anhydride or the more expensive fumaric acid are frequently used to modify rosins, particularly for use in oleoresinous media.

Frequently an esterification reaction with glycerol or pentaerythritol is used to obtain the desired final resin properties. The modifications are intended to improve the film properties of the resin but they also have an effect on the solubility and compatibility of the resins.

Theoretically only the abietic type of acids undergo maleinisation. Remembering that abietic acid isomerises to leave pimaric on heating, the following reaction is assumed to occur:

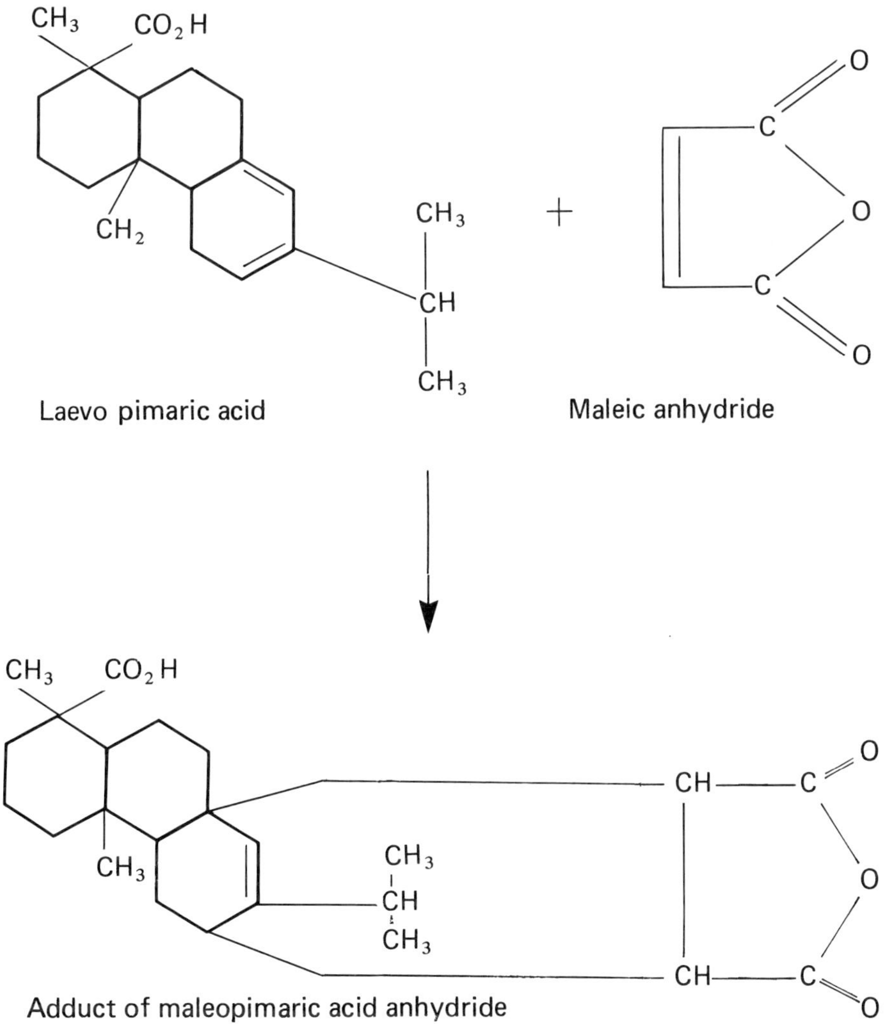

Laevo pimaric acid

Maleic anhydride

Adduct of maleopimaric acid anhydride

This reaction will readily occur at above 120° with maleic anhydride or fumaric acid. At temperatures in excess of 200°C maleic anhydride (or fumaric acid) will react with rosin via a reaction mechanism which does not involve Diels Alder adduction. It is believed (9) that some of the hydrogenated acids can undergo maleinisation to form substituted succinic anhydrides.

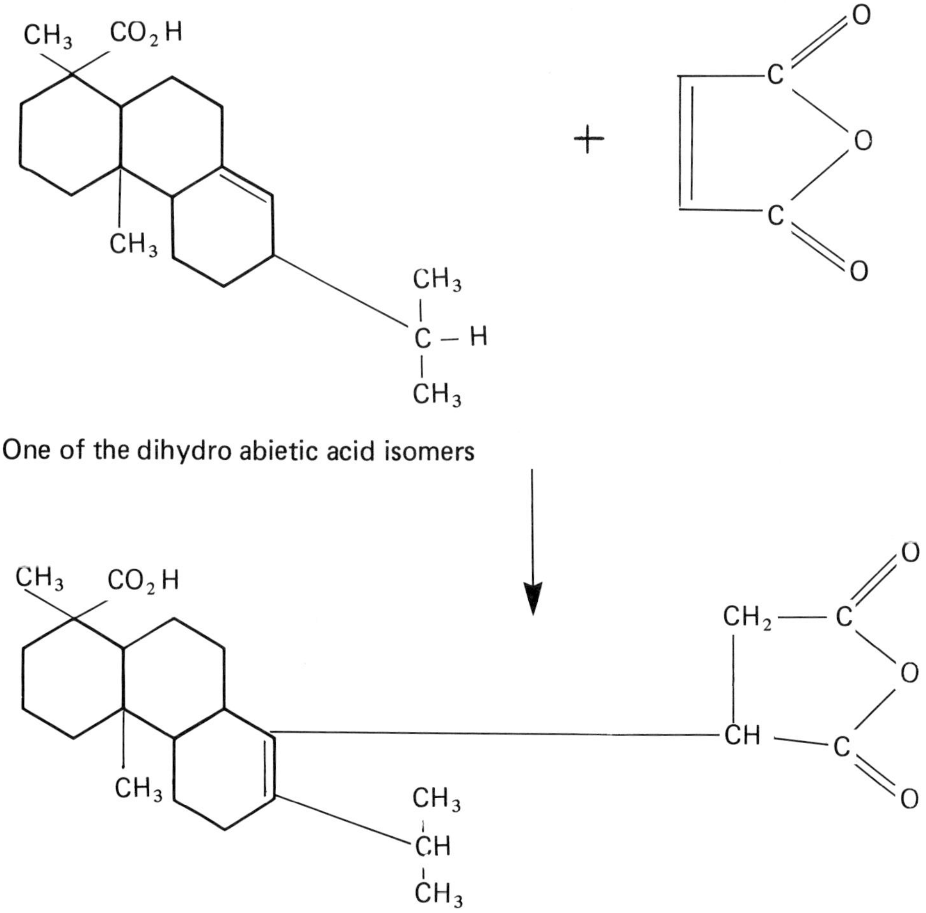

One of the dihydro abietic acid isomers

Acids other than abietic types, react with maleic anhydride as observed by Smith and Wise (9) because more maleic anhydride enters into reaction than would have been anticipated from the abietic acid content of the rosin used.

The anhydrides (or acids if fumaric is used) can readily be converted to the sodium salt by the addition of sodium hydroxide:

ROSIN MALEIC ADDUCT

Rosin	87.5
Maleic anhydride	12.5
	100.0

PROCESS

Charge the rosin and heat until it is molten (ca 90−100°C).
Maleic anhydride is added slowly and the reaction mixture
allowed to exotherm to ca 150°C. Adduction will occur at this
temperature but the temperature can be increased to 200°C if
necessary. Upon completion the resin is discharged, cooled and
kibbled.

The completion of maleinisation is indicated by determination of the unreacted maleic
content. A small proportion of free maleic anhydride does not adversely affect film
properties, particularly if the resin is to be cooked into oil, whereupon residual maleic
anhydride will react with double bonds in the oils.

Polymerised rosin can also react with maleic anhydride but because of the lack of
conjugation, higher temperatures (ca 190−210°C) are required. A lower level of maleic
anhydride is also used with polymerised rosins.

The above process may result in large loss of maleic anhydride. Several different
techniques are employed to restrict the losses during manufacture. These include charg-
ing the reactor with alternating layers of maleic anhydride and rosin and aliquot addition
of maleic to the molten rosin.

An alternative is to replace maleic anhydride by fumaric acid on an equivalent basis.
Fumaric acid is less volatile than maleic anhydride and consequently losses are
significantly less than with maleic.

Determining acid and anhydride values before, during and upon completion of reaction, will give a clear indication of the amount of maleic anhydride lost during
processing.

Maleinisation effectively increases the acid functionality of rosin from one to three. It
also raises the melting point and resistance to oxidation is improved. Gloss, light stability
and drying speed is also improved. However, there is the disadvantage of inferior
solubility in aliphatic hydrocarbon solvents.

The properties of the maleated rosin are often modified to increase its usefulness to the
surface coating chemist. A polyol, frequently glycerol, can be reacted with the adduct
to form an esterified product.

Normally rosin and maleic anhydride are charged into a reactor which is then slowly
heated to 200°C under an inert gas sparge. The maleic anhydride reacts with rosin then
glycerol is added and the reactants heated to 250°C when esterification occurs with the
anhydride and the carboxylic acid groups. Extreme care must be exercised in formulating
these systems because of the high degree of functionality.

The esters can be used in nitro cellulose lacquers. Pentaerythritol esters are harder than
the glycerol esters and have better drying and water resistance properties.

Representative formulations of a rosin maleic ester with glycerol and with pentaerythritol will be used here.

It should however, be remembered that fumaric acid can be used in any of these formulations to replace maleic anhydride provided that replacement is by equivalents.
Losses are smaller with fumaric acid than maleic anhydride.

ROSIN MALEIC GLYCEROL ESTER

Rosin	83.4
Maleic anhydride	4.6
Glycerol (10% excess)	12.0
	100.0

PROCESS

Charge the rosin and heat until molten. At about 120°C add the maleic anhydride and allow exotherm to raise temperature to 200°C. Hold at 200–220°C. Upon completion of maleinisation, as indicated by determination of the free maleic content, add the glycerol and raise the temperature to 240–260°C. Hold at 240–260°C until the acid value is below 20 mg KOH/g.

This formulation includes a 10% glycerol excess. It can be reduced for reactors which give low polyol losses. However, excess glycerol does not impair oil solubility to the same extent as excess pentaerythritol. The secondary alcohol group of glycerol is less reactive than the two primary ones so that effectively there is no free glycerol in a theoretically balanced formulation.

Often, a better control method, is to monitor acid and anhydride values during maleinisation.

This will give an early indication of any losses of maleic anhydride.

Losses may be countered by aliquot additions of rosin and maleic anhydride during maleinisation.

Two alternative processes are to mix all three components together, or to esterify the polyol and maleic anhydride, prior to addition to the rosin. Different resin companies use different techniques.

If pentaerythritol is used the hydroxyl excess required for a balance formulation should be less than 5%, otherwise the solubility in oil will be impaired.

ALCOHOL SOLUBLE ROSIN MALEIC PENTAERYTHRITOL ESTER

Rosin	83.4
Maleic anhydride	8.3
Pentaerythritol	8.3
	100.0

PROCESS

Processing would be similar to the rosin maleic glycerol ester resin, except that this resin has a high total acid value of ca 100−110 mg KOH/g, to give alcohol solubility. The melting point (B & R) is in the range 125−130°C and the resin has a viscosity of 2.5 stokes 60% non-volatile content in alcohol, ('U' tube). Alternative processes or fumaric acid can be used.

TYPICAL PROPERTIES OF
ROSIN MODIFIED MALEIC RESINS

Properties of some commercially available rosin maleic ester resins, designed for ink usage are given here to illustrate what can be obtained.

ROSIN MALEIC ADDUCT

Acid values: 115−135 mg KOH/g. Melting points 147−160°C.

Typical examples: Mitchanol 59P, 59LV, Kelrez 42-354 (10).

Recommended applications include, alcohol based gravure inks, spirit soluble overprint varnishes, aniline inks, inks for board and carton, printed circuit screen inks and some flexographic applications. The high acid value gives water dispersibility to the resin after neutralisation with amine.

WATER DISPERSIBLE
LETTERPRESS INK VARNISH

Kelrez 42−354	44−46
Tri-ethanolamine	8−10
Ethylene glycol	20−25
Di-ethylene glycol	20−25
	100

Examining some of the properties of a range of resins available from Leon Frenkel illustrates some of the trends discussed.

Rosin Type	Esterifying Polyol	Melting Point (B&R)°C	Acid Value mg KOH/g	White Spirit Tolerance
Rosin	Glycerol	84 – 87	6	1 : 10
Rosin	Glycerol	84 – 88	10	1 : 10
Rosin	Pentaerythritol	95 – 105	10	1 : 10
Maleic rosin	Glycerol	95 – 100	15	1 : 10
Maleic rosin	Glycerol	115 – 125	15	1 : 10
Maleic rosin	Glycerol	129 – 132	25	< 1 : 1
Maleic rosin	Glycerol	135 – 145	25	< 1 : 1
Maleic rosin	Glycerol	150 – 155	25	< 1 : 1
Maleic rosin	Pentaerythritol	123 – 127	10	1 : 10
Maleic rosin	Pentaerythritol	143 – 147	12	1 : 2
Maleic polymerised rosin	Pentaerythritol	120 – 125	15	1 : 10
Maleic polymerised rosin	Pentaerythritol	140 – 145	15	1 : 10
Maleic polymerised rosin	Pentaerythritol	153 – 157	15	1 : 5
Fumaric rosin alcohol soluble		140 – 150	280	Insoluble
Fumaric rosin alcohol soluble		125 – 130	260	Limited
Maleic rosin alcohol soluble	Pentaerythritol	140 – 155	120	Insoluble

As the hardness (i.e. melting point) of the ester increases, solubility in white spirit decreases, for both glycerol and pentaerythritol esters.

A maleic rosin ester is harder than the corresponding rosin ester. Maleic polymerised rosin esters, tend to be high melting point resins. The replacement of maleic anhydride by fumaric acid is illustrated. Alcohol solubility is introduced by retaining a high acid value.

ROSIN MALEIC ESTER RESINS
Melting point and viscosity for rosin maleic ester resins can be easily varied by the selection of esterifying polyol, by the amount of polyol used and by the degree of reaction. They find use in oleoresinous media, including cold cut varnishes, lithographic and letterpress media and overprint varnishes.

Typical properties of glycerol esterified resin, would be acid values of 15−30 mg KOH/g and melting points of 120−140°C.

Whilst pentaerythritol esterified resins may have similar acid values and melting points, their films are much harder than they are for an equivalent degree of reaction for a glycerol based resin.

For a rosin ester with an acid value less than 20 mg KOH/g, the maximum amount of maleic anhydride which can be incorporated, is 9% of the rosin content. Further quantities will result in gelation unless the reaction is stopped at a higher acid value.

As the maleic anhydride content increases the melting point increases. Compatibility however, tends to decrease.

The following tables, some of which are reproduced from the OCCA book (II) illustrate many of the important trends.

FOR A 10% PENTAERYTHRITOL ESTER

Maleic Content % in Rosin	Acid Value mg KOH/g	Softening Point (B&R)°C
3	23	123
5	23	129
6	23	134
9	21	152

FOR A GLYCEROL ESTERIFIED MALEIC RESIN (II)

Maleic Anhydride %	Acid Value mg KOH/g	Melting Point (B&R)°C	Dilution in White Spirit	Solubility in 50 poise Dehydrated Castor Oil
0	8	85	1:20	Easily
4	15	105	1:5	Easily
8	15−20	125	1:2	Easily
12	25	136	1:1	Difficult cooking required

EFFECT OF GLYCEROL AND PENTAERYTHRITOL ON THE PROPERTIES OF ROSIN MALEIC ESTERS (II)

Polyol	Maleic Anhydride %	Melting Point (B&R)°C	Solubility in White Spirit and Drying Oils	Compatibility with Nitro Cellulose
Glycerol	5	110	Easily soluble	Compatible
Pentaerythritol	5	125	Easily soluble	Poor
Glycerol	7½	125	Less soluble	Compatible

As can be seen above, increasing the maleic anhydride content to about 12% for a glycerol modified ester, makes it very difficult to obtain solubility in oil at low concentrations of resin.

As would be expected of a tri-functional system, increasing the maleic anhydride increases the molecular weight of the resin and the complexity of the molecule.

Care must be exercised when forming oleoresinous media, to avoid further condensation and subsequent gelation.

Maleic rosin esters have superior non-yellowing properties to ester gums and more importantly they do not darken during the formation of an oleoresinous medium. These esters can be used with dehydrated castor oil to give relatively low colour oleoresinous varnishes. The lower maleic anhydride content pentaerythritol rosin maleic esters are not as resistant to yellowing as the glycerol esters.

Maleic resins are classified as soft or hard. Soft resins are easily dispersed in drying oils at 250°C.

Hard resins require solubilising, with not more than an equal weight of oil at 250°C until a clear bead is obtained. Then small quantities of additional oil are progressively cooked in, until the required concentration is reached.

As can be seen from the tables, increasing maleic content decreases solubility in white spirit. Alcohol solubility can be improved by using a polyol excess, but oil and hydrocarbon solubility may suffer.

Maleic resins are frequently used with nitrocellulose to give gloss and adhesion. Maleic rosin esters are compared against rosin modified phenolic resins in a later section of this chapter.

Hydrogenation

Hydrogenation can be carried out at high pressures using a nickel catalyst. The products of hydrogenation were considered under the disproportionation section.

Catalysts can be easily poisoned by sulphur in the rosin. Hydrogenation improves resistance to oxidation and the products are used to prepare light coloured rosin esters of glycerol, pentaerythritol and ethylene glycol.

ROSIN MODIFIED PHENOLIC RESINS

Phenolic resins are the subject of a separate chapter. However, there is one class of phenolics which are dealt with in this section. They are the rosin modified phenolics.

Phenolic resins are divided into two major types: i) Novalacs; and ii) Resoles.

Novalacs

Novalacs, which are thermoplastic, find limited use in the majority of surface coating and ink formulations because of their insolubility in oils and white spirit. They can be used however, for specialist or novel applications.

Novalacs are formed from using a formaldehyde/phenol ratio less than unity. The simplest representation of a novalac resin is:

However in practice there are many repeat units.

Resoles

If however, an alkaline catalyst like lime or NaOH is used, a reactive phenolic called a resole is obtained. The formaldehyde/phenol (f/p) ratio used is greater than one and upon heating a resole, further condensation can occur.

The chain structure contains reactive residual methylol groups (CH_2OH). The simplest representation of a resole resin is:

However, in practice the structure is more complicated, particularly as the formaldehyde can react in the para position as well. This gives rise to their cross linking and further reaction upon heating. Residual methylol groups are very reactive.

Phenolic resins are not limited to phenol. Substituted phenols are frequently used in surface coating applications. Examples would be:

| p-tertiary butyl phenol | Alkyl phenol (e.g. octyl) | di-phenylol propane |

Alternatively, cresols which consist of a mixture of isomers with the meta form being the most reactive, are usable:

| meta cresol | ortho cresol | para cresol |

As a general rule both types of (phenol) phenolic resins find little use in surface coatings and inks without being modified.

Novalacs can be alcohol soluble. Initially they were merely made oil soluble by incorporating into synthetic copals. Film properties and film performance were poor, however, as was colour. They tended to be used for dark tung-based oleoresinous media only.

However, the incorporation of a small amount of a resole phenolic into rosin, vastly improves the film properties and induces oil solubility. The resole is cooked into rosin or more normally into rosin ester.

Normally oil solubility is induced by cooking a small proportion (5−20%) of a resole phenolic, into rosin between 160−270°C (normally ca 250−270°C). A polyol can then be added to reduce acid value and improve performance.

The formation of an oil soluble resin enables a wide range of oleoresinous media to be prepared and oil solubility can be introduced into phenolic resins by using large para substituents on the phenyl group of the phenolic.

Obviously, varying the substituents will also affect oil solubility and other properties in a rosin phenolic resin and a combination of effects can be used. Cresylics and di-phenylol propane (DPP) based phenolics can also be modified by rosins.

To differentiate between oil soluble rosin modified phenolics and other oil soluble phenolics, the latter are termed 100% phenolics (either novalac or resole).

The chemical reactions between rosin and resole have not been fully elucidated. Reaction is believed to occur between the double bond in rosin acid and a methylol group in the resole. The simplest reaction scheme is (11).

An alternative reaction summarised in Solomon (12) gives rise to chroman ring formation which is believed to enhance oil solubility.

Reaction between the acid group of rosin and the hydroxyl (methylol) groups is theoretically possible, but the acid value change, with due allowance for decarboxylation, indicates that there is little or no esterification between them (11), (13).

A simple comparison of the molecular weights of phenol, formaldehyde and rosin illustrates one of the reasons for incorporating a relatively low level of resole into rosin.

The smallest molecular weight resole has a weight of 154 and rosin 300. For two mols to one of resole there is a 4:1 ratio. This means that 20−25% of a resole will react with most of the unsaturated acids present in rosin leaving very little free acid.

The effects of phenol formaldehyde ratios and other variable parameters on the properties and performance of these resins are discussed in the section on general properties.

As was the case with rosin maleic resins, the nature and level of the esterifying polyol also has a considerable effect upon properties.

The Preparation of
Rosin Modified Phenolic Resins

The normal method of preparing rosin modified phenolics is to slowly add the resole to molten rosin (say between 150−210°C). An inert gas atmosphere is used. Upon completion of the resole addition, the polyol is added and the temperature is raised to 240−270°C. The reaction is monitored by acid value decrease and melting point increase and this indicates completion of reaction. The final phenolic esterified rosin resin normally contains only 7−12% resole.

Below are three starting formulations for rosin modified phenolics and each utilises a different type of phenol. The melting points of all these resins are in the region of 135−140°C. The acid values are below 20 mg KOH/g for the first two examples and 25 mg KOH/g for the third.

SIMPLE RESOLE MODIFIED ROSIN

Rosin	80.0
Phenolic (resole)	11.2
Pentaerythritol	8.8
	100.0

DI-PHENYLOL PROPANE BASED RESOLE

Rosin	80.0
DPP phenolic	11.2
Pentaerythritol	8.8
	100.0

The use of a magnesium oxide (0.1%) catalyst is optional.

p-t BUTYL PHENOLIC MALEIC ROSIN ESTER

Rosin	80.3
Maleic anhydride	4.6
Pentaerythritol	10.1
p-t Butyl phenolic	5.0
	100.0

General Properties of Rosin Phenolics

Rosin phenolics are generally hard, brittle resins with melting points ranging from 100–160°C. Esterified rosin phenolics tend to have acid values in the region 10–25 mg KOH/g whilst the unesterified ones are 100–140 mg KOH/g.

Varnishes based on rosin phenolics are faster drying, have better durability, are harder and have greater resistance to water than ones based on ester gums. However, their colour stability is inferior and this is more noticeable than with some other resins because the initial colour of the phenolic resins is paler.

Phenolic resins were classified as hard or soft and the soft types with melting points in the region of 120°C readily dissolve in white spirit. These have a high tolerance to dilution, unlike hard types.

Hard types with melting points in the region of 150°C are difficult to dissolve and often precipitate if diluted further.

However, the resins either soft or hard have good compatibility with glyceride oils, even though some degree of cooking may be required. (Unlike natural resins, it is not necessary to 'run' this class of phenolics with associated weight losses, to obtain oil solubility.)

Modified phenolics can have a retarding effect on the gelation of wood oil. Modified phenolic resins can be used in media for paint (gloss), varnishes, primers, overprint varnishes, litho, letterpress and roto gravure inks. Substituted phenols (e.g. p-t butyl phenol) are used to reduce yellowing.

Esterified rosin phenolics are generally insoluble in alcohols and are soluble in aromatic hydrocarbons, white spirit and glyceride oils. Where alcohol solubility and solubility in hydrocarbons and oils is required, rosin phenolics are normally only partially esterified to leave a high acid value.

Typical phenolic resoles used to modify rosin esters have formaldehyde, phenol ratios (F/P) ranging between 1.6:1 to 1.8:1 and may be based upon phenol, p-t butyl phenol or cresols. If di-phenylol propane is used the F/P ratio is 2.5:1 or higher. The following tables illustrates some of the variations in properties with F/P ratio.

F/P Ratio	% Phenolic	Type of Phenolic	Acid Value mg KOH/g	Melting Point (B&R)°C
1.8:1	5	p-t Butyl phenol	20	130
2.5:1	7	D.P.P.	20	130
3.5:1	7	D.P.P.	15	135
3.5:1	8	D.P.P.	15	140

Increasing the level of phenolic resole in the rosin ester, increases the melting point, reduces oil compatibility and solubility in white spirit. This is illustrated by referring to a table in the OCCA paint technology manual (ref. (11) p. 64).

Level of Resole %	Melting Point (B&R) °C	Acid Value mg KOH/g	Solubility in White Spirit	Solubility in Linseed Stand Oil
0	85	10	All proportions	Very easily
10	120	15 – 20	1:5	1:2 at 200°C
20	140	20	1:1	1:1 at 260°C
25	150	20	Barely 1:1	1:1 at 260°C

Some of the properties of a range of rosin modified phenolic resins illustrate how they can vary with type of phenol. All except the high acid value ones are esterified by glycerol. All of these resins are available from the Cornbrook Resin Company.

Type of Phenol	Acid Value mg KOH/g	Melting Point (B&R)°C	Alcohols	Solubility Aromatics	White Spirit	Drying Oil
D.P.P.	15 – 20	115 – 125	I	S	1:3	Excellent
D.P.P.	15 – 20	135 – 140	I	S	1:2:5	Good
D.P.P.	15 – 20	140 – 150	I	S	I	Oil up technique
Cresylic	135 – 145	100 – 115	S	S	L	Very good
Cresylic	120 – 130	120 – 125	L	S	L	Very good
Cresylic	15 – 20	115 – 125	I	S	1:4	Excellent
Cresylic	15 – 20	125 – 135	I	S	1:3	Excellent
Cresylic	15 – 20	140 – 150	I	S	I	Oil up technique
p-alkyl phenol	15 – 20	115 – 125	I	S	S	Excellent
p-alkyl phenol	15 – 20	135 – 145	I	S	S	Excellent
p-alkyl phenol	15 – 20	145 – 150	I	S	1:4	Excellent

I = Insoluble   S = Soluble   L = Limited

The change in solubility with increasing hardness (i.e. melting point) of the resins are obvious, as is the change in alcohol solubility with unesterified resins.

The effects of esterifying alcohol can be illustrated by examining a range of resins obtainable from Leon Frenkel. Again increasing hardness decreases tolerance for white spirit.

Esterifying Polyol	Melting Point (B&R)°C	Acid Value mg KOH/g	White Spirit Tolerance	
Glycerol	115 – 120	10	1:10	
Glycerol	120 – 130	25	1:10	
Glycerol	125 – 130	10	1:10	
Glycerol (Cresylic)	135 – 145	15	1:1	
Pentaerythritol	125 – 130	10	1:10	
Pentaerythritol	135 – 140	10	1:10	
Pentaerythritol	150 – 160	30	1:5	
Nil	120 – 125	130	1:2	alcohol soluble
Nil	125 – 135	100	1:10	alcohol soluble

Some starting formulations, published in resin suppliers literature for inks and varnishes containing rosin phenolic modified resins are given below.

Formulations

WEB O/S HEATSET VARNISH (LOW TEMP)
(Warwick International Ltd)

Mitchanol 42*	43.0
Surkyd 127*	8.5
Pkw F4/7 af (New) (Halterman)	32.0

Heat to 180°C hold until dissolved. Cool to 100°C and add:

Monalon 205 (Man chem)	0.5
Pkw F4/7	9.5

Premix and add. Heat to 125°C and hold for ½ hour. Cool and add:

Pkw F4/7	5.5
Tridecon1	1.0
	100.0

*Available from Warwick International Ltd.

QUICKSET O/S VARNISH
(Warwick International Ltd)

Mitchanol 42*	45
Surkyd 408*	10
Linseed oil alkali refined	25

Heat to 200°C. Hold until dissolved. Cool and add:

260 – 290°C PID	20
	100

WEB O/S HEATSET INK LOW TEMP
(Warwick International Ltd)

Blue pigment	16.0
Heatset varnish	65.0
Wax (solids)	1.5
Hakuerka OT (Croxton & Garry)	3.5
Grind and add:	
Pkw F 4/7	11.5
Pkw F 4/7	2.5
	100.0

QUICKSET O/S INK
(Warwick International Ltd)

Blue pigment	16.0
Hakuerka OT	4.0
Quickset varnish	62.0
Wax (solids)	2.0
Grind and add:	
280 – 310°C PID	12.5
Cobalt (10%)	0.5
Manganese (10%)	0.5
280 – 310°C PID	2.1
	100.0

OFFSET GLOSS INK (Reichhold)

*Flush colour		40 – 60
**Varnish		20 – 40
100% Non-volatile content long oil alkyd	(see below)	10 – 20
Polyethylene compound	(see below)	5 – 7
6% Manganese drier		1 – 2
6% Cobalt drier		½ – 1
Magie 535		3 – 5
		100

*RD-985	Resinated Di-arylide Yellow	— Ridgway Colour
3LB-443	Phthalocyanine Blue	— Pope Chemical
528G-106	Litho Rubine	— Hilton Davis

Gloss Varnish	or	**Quick Set Varnish**	
24 – 110	– 40	24 – 220	– 20
A.R.L.O.	– 60	24 – 105	– 20
	100	A.R.L.O.	– 25
		Magie 535	– 35
			100
Visc = Z_{10} @ 25°C		Visc = Z_9 @ 25°C	

N.B. *Either varnish may be used as the sole vehicle, or combinations of the two may be employed to impart desired gloss and setting characteristics.*

OFFSET QUICKSET INK (Reichhold)

*Quickset flush	(see below)	50.0
**Quickset varnish	(see below)	38.0
Polyethylene wax compound		7.0
6% Manganese drier		1.0
6% Cobalt drier		.5
Magie 535 oil		3.5
		100.0

Quickset Flush (for above offset ink)

528K-125	Resinated Litho Rubine — Hilton Davis
527K-297	Resinated Di-arylide Yellow — Hilton Davis
3-LB-443	Phthalocyanine Blue — Pope Chemical
FS-926	Macatawa Red (Rubine) — Chemetron
RD-985	Resinated Di-arylide — Ridgway Colour & Chemical
RD-984	Rhodamine Y PTMA — Ridgway Colour & Chemical
RD-987	Phthalocyanine Blue — Ridgway Colour & Chemical

Quickset Varnish (for above offset ink)		
**24 – 109	40 to	45
A.R.L.O.	30 to	20
Magie 535	30 to	35
	100	100

SOME MISCELLANEOUS REACTIONS OF ROSIN

Rosin Formaldehyde: Rosin will react with formaldehyde, forming an adduct, with reduced tendency for crystallisation. It can be used in some inks.

Rosin/Reactive Hydrocarbon Resin: Reactive hydrocarbons like Escorez 8000 will react with wood rosin at 260°C, with a reduction in acid value, and increase in melting point.

NATURAL RESIN MEDIA AND VARNISHES

Many of the natural resins require modification before they can be used in a coating system. Natural resins were the first resins to be used in the surface coating industry and they were processed in open pots or kettles which were directly 'fired' (i.e. heated) by gas. They resembled large scale bunsen burners and crucibles. Extreme care was required to stop the contents of the pot igniting, particularly when foaming occurred. Fires were an everyday hazard for the early varnish makers.

Varnishes, which are translucent liquids, 'dry' (i.e. harden) by solvent evaporation, or by chemical cross linking, caused by exposure to oxygen in the air. The latter, normally referred to as oleoresinous varnishes, are considered in a later section of this Chapter.

Controls for resin during processing, are normally very simple and relate to the requirements of the resin. The normal ones are viscosity, compatibility with an oil, or other resin (e.g. nitro cellulose), solubility in an oil or solvent, or a bead of the reaction mixture clearing on cooling to indicate a complete reaction.

It often is necessary to 'cook' the system, until it is on the verge of 'gelling' to obtain maximum performance from the coating or ink. This is particularly true for short oil systems. The normal method of indicating the completion of the reaction, was to cause the resin to form a string and measure the length of the string. A string can be formed either hot by allowing resin to run off a spatula, or cold by pulling a cooled film between the fingers or from a cooled sample.

It is essential to rapidly cool the reaction mixture upon completion of reaction, otherwise gelation of the contents of the reaction vessel may occur. Normally rapid cooling is achieved by adding further cold oil. Varnishes can also be used as media for inks or paints.

Before considering the modification of natural resins into varnishes, one of the oldest varnishes will be briefly discussed. Linseed Oil (see Chapter II) is a drying oil and its drying can be improved by heat treating. Some of these treatments were discussed in Chapter II.

Litho varnishes can be prepared by heating linseed oil, in an open pot at 250−290°C, with access to air. Acidity is carefully controlled, to a value between 5−15mg KOH/g, otherwise excessive emulsification may occur during printing.

Inert gases can be 'blown' (sparged) through the oil during processing to remove excess acid. However, increasing viscosity, which indicates increasing degree of polymerisation, is characterised by increasing acid value. This can be overcome by heating the oil, in a closed pot, under vacuum or with an inert gas sparge. The colour is normally much improved, as is the odour, because most oxidation products have characteristic smells.

Boiled oil is obtained by processing between 95−150°C in intimate contact with air and in the presence of driers. This should increase the speed of drying without adversely affecting viscosity. Again the oil is partially polymerised so that the number of cross links required to form a coherent tack-free film is reduced. The viscosity is lower than equivalent heat treated oils, because it is less heat polymerised than a litho oil and less oxidised than a 'blown' oil.

To improve the solubility of many natural resins it is normal to 'run' them, i.e. effectively to depolymerise them. This is normally accompanied by fuming and sometimes the heat history is important and for each source or batch of natural resin, it may be necessary to modify the process slightly due to differing compositions.

A congo copal resin would be run at 320−360°C, with a weight loss of ca 25%. Control would be by solubility and preheated linseed oil could be 'cooked' in, after running the copal, to form an oleoresinous media. The lower the temperature of running, the paler the product, but the higher the viscosity. Air must be excluded during 'running'. The lower temperature product is a slack melt and requires removal of surface scum, whereas the higher temperature fine melt does not. The molten resin is collected in trays and crushed when cold. The effect of running on the properties of a congo copal would be (14).

	Before Running	After Running
Melting point (B&R) °C	120−220	60−120
Acid value mg KOH/g	95−115	50−115
Iodine value	110−130	110−125
Saponification value mg KOH/g	120−150	55−140
S.G.	1.05	1.03

The acid value of run copal can be reduced by reaction with lime or glycerol. As can be seen the process of running has caused (14).

  i) Decarboxylation of di-basic acids to mono-basic acids;

 ii) Conversion of hydroxy acids to lactones;

iii) Decomposition of higher mono-basic acids into hydrocarbons and lower acids.

These three reactions can be summarised as follows:

$$i) \; HO_2C{-}R \; CO_2H \; \xrightarrow[heat]{} \; HRCO_2H + CO_2$$

$$ii) \quad \begin{array}{l} HC - CO_2H \\ \quad | \\ HC - OH \\ \quad | \end{array} \quad \xrightarrow{-H_2O} \quad \begin{array}{l} HC - C = O \\ \quad | \quad\quad | \\ HC - O \\ \quad | \end{array}$$

on further heating the lactone may decompose to a ketone:

$$\begin{array}{l} \quad | \\ HC - C = O \\ \quad | \quad\quad | \\ HC - O \\ \quad | \\ \text{Lactone} \end{array} \quad \xrightarrow[-CO]{} \quad \begin{array}{l} \quad | \\ HCH \\ \quad | \\ C = O \\ \quad | \\ \text{Ketone} \end{array}$$

$$iii) \; HRCO_2H \; \xrightarrow[heat]{} \; HRH + CO_2$$

There are alternative methods of making copal oil soluble and these include (14).

i) Mastication by heated rollers between 80−200°C. This can even cause alcohol solubility;

ii) Heat treatment with phenol or cresol using tetra hydronaphthalene as a solvent;

iii) Crack the resin using hydrogen and a Raney nickel catalyst.

Upon the completion of 'running', the resin can be either cooled and dissolved in solvent (e.g. white spirit) or dispersed (solubilised) in oil. Alternatively oil can be cooked with the resin. The former are called 'cold cut' varnishes or media because only physical mixing occurs. With the latter there is some degree of chemical reaction which is often ester interchange.

Cold cut systems are often used during ball milling where the milling assists the solubility of the resin in solvent and oil (if present). However, the properties of a cold blend are often considered inferior to an 'equivalent composition' which has been 'cooked in'. When oil is present with resin the system is termed oleoresinous.

The acid constituents of natural resins are able to undergo many of the reactions of carboxylic acids, with esterification with a di, tri or tetra polyol being the commonest.

CONGO COPAL ESTER

Formulation	
i) Copal	94.34
ii) Glycerol	5.66
	100.00

PROCESS

Charge kibbled (i) and heat to 300°C. Hold at 300°C for 15 mins. Cool to 290°C and add (ii). Hold at 290°C until completion of reaction which will take 10—15 hours and is indicated by acid values and solubility in white spirit. When solubility is acceptable, cool. If required, white spirit can then be added.

MODIFIED CONGO COPAL ESTER

Formulation	
i) Copal	83.95
Rosin	9.33
ii) Glycerol	6.72
	100.00

PROCESS

Process as above. Note that the amount of glycerol is adjusted because rosin requires about twice as much glycerol for esterification as copal.

HARD COPAL/LINSEED OIL VARNISH

Formulation	
i) Hard copal	47.45
ii) Linseed oil	35.55
iii) White spirit	17.00
	100.00

PROCESS

Charge (i) and heat to 300°C determining solubility in oil as described earlier. A 25% weight loss should have occured when running should be complete. Cool to 275°C and add (ii) preheated to 160°C. Continue heating until a drop clears upon cooling whereupon cool to 160°C and charge (iii).

OLEORESINOUS VARNISHES AND MEDIA

The three major constituents of an oleoresinous varnish are:

i) **Resin:** either natural or synthetic to impart gloss, hardness, durability and non-blocking.

ii) **Drying Oil:** for drying on exposure to air to complete cross linking of the film and improve film properties.

iii) **Volatile Solvent:** for adjusting the application viscosity to suit the application method (i.e. brush or spray) or printing press to be used.

Driers, anti-skin agents, etc. would be added to adjust rate of drying. If this varnish is to be used for an ink or paint media then pigment, etc. would be dispersed in it and further oil and varnish added.

Oils have been considered in detail in Chapter II. The oils which find usage in oleoresinous media are normally drying, but semi-drying ones are said (13) to work.

One of the most popular is, and was, linseed oil in one of its many forms (e.g. stand oil), but china wood and esterified tall oil can be used. Dehydrated castor oil has limited uses. Tung oil, a quicker drying oil can be incorporated into many lacquers and varnishes. Soya is claimed (13) can be used, but the drying characteristics are poor.

The oil content of an oleoresinous mixture is nowadays described by 'oil length', which has the same meaning as that described in the Alkyd Chapter (Chapter IV).

Linseed oil, in one of its forms, was used for many years, but it was found that the incorporation of the faster drying tung oil improved the properties of the oleoresinous varnish. Initial attempts at incorporating tung oil were unsuccessful with gelation resulting, but incorporating rosin or a phenolic resin avoided this problem. With oil (and tung in particular), it is necessary to carefully control batches, on a time-temperature basis and use rapid cooling which is normally achieved by the addition of linseed or alternative oil. Gas checking discussed in Chapter II is a problem with tung oil.

It is possible to prepare phenolic resins which are oil soluble without modifying by rosin. Oil solubility can be conferred by the size and nature of the hydrocarbon constituent in the phenolic resin. Phenolic resins may be used with tung oil to give excellent films of good water and chemical resistance, flexibility and rapid drying. The major drawback is yellowing. Phenolic oleoresinous systems can be used for varnishes.

For resoles, oil solubility is induced by:

i) Increasing the ratio of formaldehyde to phenol so that it is greater than 1.

ii) Leaving only two reactive positions of the phenyl nucleus unsubstituted.

iii) Only one alkyl group containing three or more carbon atoms or an aromatic group should be in the para position.

These resins have a low melting point which increases on heat processing. The reaction which occurs between the resole and oil have not yet been fully elucidated. Resoles are preferred to novalacs in most oleoresinous systems.

With the solvent and use of phenolic resins which provided a source of synthetic resin with more consistent properties, the less skilled varnish maker could relatively easily manufacture oleoresinous media. Many of the variations in properties of the media which resulted from variations in the natural resins were eliminated. Phenol resins enabled oleoresinous media to be obtained with consistent properties.

As will be appreciated, the ink industry is particularly secretive and sensitive about the actual composition and methods used in manufacturing oleoresinous media. Therefore all formulations will be quoted from the literature (3, 4, 10, 11, 14, 15, 16) and formulators must consider these as guidelines or starting formulations which need to be modified to suit their requirements and materials.

It is not possible to produce pale oleoresinous varnishes by bodying the oil and resin together, thus stand oils are frequently used. To enable control, during processing, it is necessary to determine when the resin becomes soluble in stand oil. To determine solubility, heat a 50/50 mixture of resin and stand oil (100 poise), until a clear bead is obtained when a small sample of the mixture is cooled on a piece of glass. Three parts of pre-heated stand oil are added slowly, followed by petroleum thinners. If the solution remains perfectly clear when cooled, then the resin is soluble in stand oil at that temperature.

It is also necessary to determine what viscosity of oil is required to give the desired 'body'. As a general rule, resins which become soluble in oil at comparatively low temperatures, have low body and require a more viscous stand oil than the less soluble varieties. For any given resin, the longer the oil length required, the higher the viscosity of the stand oil necessary to achieve this.

As a guide for processing temperatures (16) fused Congo copal requires heating to 300°C with stand oil, Congo ester 260°C and synthetic copals 200−285°C, before solubility is attained.

Short oil, low non-volatile content varnishes can be made by heating the oil at 300°C until it starts to become stringy, adding litharge to prevent gelation and then cooking the resin in.

It is possible to cook many types of resin into drying oils. There are many hydrocarbon resins commercially available. Generally speaking the resin maker purchases these resins because of the lack of feedstock (often naphtha) and sophisticated high tonnage plant required for cost effective polymerisation. Typical examples taken from supplies literature have been given here to give the reader a feel for formulations and processing conditions.

Also quoted, is an example of a hydrocarbon resin (Imprez), being cooked into an alkyd. Some examples illustrating preparations of oleoresinous media will be given.

REACTIVE HYDROCARBON RESIN/LINSEED OIL

Formulation	
Escorez 8000	30 – 40
Linseed oil	70 – 60
	100

PROCESS

Charge Escorez 8000, which is a reactive hydrocarbon resin with a high degree of unsaturation (available from Esso), (4) and linseed oil. Heat to 260°C under reflux. Smoke will appear during reflux. The temperature will slowly rise from 260°C to 290°C. Hold at 290°C until the desired viscosity is attained and the resin which is liquid and insoluble in acetone, does not show any signs of phase separation between oil and Escorez.

The ratio of Escorez/oil is selected to suit the viscosity required, but at a ratio greater than 40/60, reproducible properties cannot be obtained on a production scale.

To stop a rapid viscosity increase once the bottom of the viscosity range is reached, rapid cooling is required and this can be enhanced by the addition of white spirit. The Escorez/oil copolymers are compatible with stand oils and long oil alkyds and exhibit quicker drying with improved hardness.

VEHICLES FOR LETTERPRESS PRINTING INKS

Using Resen 105−115, which is a 100% hydrocarbon resin (aliphatic and aromatic) from Albright & Wilson, a variety of printing ink vehicles for offset inks can be prepared.

Low Cost Ink Vehicle	
Resen 105 − 115	40
Mineral oil	60
	100

Process at 120 − 130°

Medium Cost Ink Vehicle	
Resen 105 − 115	40
Linseed oil	30
Mineral oil	30
	100

Process at 120 − 130°

High Quality High Cost Ink Vehicle	
Resen 105 – 115	30
Other resin (e.g. phenolic or rosin)	10
Mineral oil	25
Linseed oil	35
	100

Process at 165 – 175°

2% of (6%) cobalt octoate or
naphthenate can be incorporated as driers.

MODIFIED PHENOLIC LINSEED OIL
GLOSS INK AND OVERPRINT VARNISH

Amberol F-71	86.21
Litho oil No. 00	13.79
	100.00

PROCESS

Charge and slowly heat to 205°C. Hold for 2 hours at 205°C or
until bead clears. Cool. Amberol F-71 is a modified phenol formaldehyde resin available from Rohm and Haas.

ROSIN MODIFIED MALEIC
OLEORESINOUS INK VEHICLE

i) Amberol 801	72.46
Dehydrated castor oil	7.25
ii) Mineral oil	20.29
	100.00

PROCESS

Charge Amberol 801 which is a rosin modified maleic resin and
DCO and heat to 305 – 310°C until a clear bead is formed. Cool
to 120°C and add ii).

A CONGO COPAL ESTER VARNISH FOR A
GENERAL PURPOSE GLOSS PAINT VEHICLE

i)	Congo ester	12.09
	Linseed stand oil (100 poise) (A)	12.09
ii)	Tung oil	9.07
	Linseed stand oil (B)	27.20
iii)	Litharge	0.48
iv)	Mineral spirit	38.68
	6% Cobalt naphthenate	0.39
		100.00

PROCESS

Charge i) and heat to 260°C. When a bead of the mixture clears add ii) and continue to hold the temperature at 260°C for 40 minutes then cool to 230°C over 20 minutes and add iii). Continue to cool to 180°C over 30 minutes and then add iv). The viscosity is about 2–3 poise (25°C) and the drying time 16–18 hours at 15.5°C.

MODIFIED PHENOLIC TUNG OIL VARNISH
FOR AN INTERIOR GLOSS PAINT VEHICLE

i)	Phenolic resin	18.05
	Tung oil	15.04
	Linseed stand oil (30 poise)	15.04
ii)	Linseed stand oil (100 poise)	15.04
iii)	Mineral spirit	36.10
	6% Cobalt naphthenate	0.73
		100.00

PROCESS

Charge i) and heat to 285°C over 100 mins. Charge ii) and allow to 'chill', i.e. (cool to 230°C over 5 minutes). Allow gradual cooling from 230°C to 180°C over 40 minutes and then add iii) cool and discharge. The viscosity is 2 to 2.5 poise (25°C) and the drying time at 15.5°C is 4–6 hours.

A CHEAP ROSIN ESTER TUNG OIL VARNISH

i)	Wood rosin	14.96
	Pentaerythrital	1.40
	Tung oil (A)	1.87
ii)	Tung oil (B)	42.99
iii)	Mineral spirit	37.38
	6% Cobalt naphthenate	1.40
		100.00

PROCESS

Charge i) and heat to 285°C. Hold at 285°C for ca 1 hour and then add ii). Reheat to 285°C. Cool to 180°C over 1 hour and add iii). The viscosity is about 2.5 poise (25°C) and the drying time is 16 hours at 15.5°C.

TRAFFIC PAINT BASED ON
AN OLEORESINOUS VEHICLE

Formulation of Oleoresinous Vehicle

i)	China wood oil (or other drying oil)	31.36
	Resen (A)	7.84
ii)	Resen (B)	7.84
iii)	Aliphatic solvent	25.78
	Aromatic solvent	25.78
iv)	Lead naphthenate (24%)	0.70
	Cobalt naphthenate (6%)	0.70
		100.00

PROCESS

Charge i) and heat to 290°C. Hold for 15 minutes. Cool and add ii). Reheat to 290°C and hold at 290°C until desired viscosity is attained. Cool, add iii) and iv).

Formulation of Paint	
Oleoresinous vehicle (as above)	28.49
Chlorinated rubber	3.41
Aromatic solvent	3.41
Titanium di-oxide	11.53
Calcium carbonate	41.12
Barytes	6.52
Mica	5.52
	100.00

PROCESS

Dissolve the chlorinated rubber in solvent and disperse pigment in the normal manner. A cold cut version can be prepared by mixing boiled linseed oil and Resen 95–105 with pigments extender and solvent.

WATER RESISTANT SPAR VARNISH

Formulation	
i) Phenolic resin	2.33
Refined linseed oil	9.33
ii) Tung oil	9.33
iii) Resen 95 – 105	56.68
iv) Mineral spirits	22.00
Lead naphthenate (24%)	0.22
Cobalt naphthenate (6%)	0.10
Anti-skin agent	0.01
	100.00

PROCESS

Charge i) and heat to 305°C over 35–40 minutes. Hold at 260°C until desired viscosity is attained. Add iii) rapidly, then iv) and cool. The drying time is 6–8 hours and is unaffected by 72 hours exposure to water.

ALKYD VEHICLE

i)	Resen 95 – 105	8.35
	Linseed oil acids	26.16
	Glycerine	10.58
	Phthalic anhydride	15.86
ii)	Xylene	2.23
iii)	Mineral spirits	35.63
	Lead naphthenate (24%)	0.61
	Cobalt naphthenate (6%)	0.56
		100.00

PROCESS

Charge i) and process as an alkyd at up to 250°C. When required add ii). Hold for acid value less than 5mg KOH/g. Cool and add iii).

COLD CUT AIR DRYING ALKYD PAINT

i)	Alkyd (A)	35.68
	Black pigment	1.78
ii)	Alkyd (B)	35.68
iii)	Resen 95 – 105	8.33
	Mineral spirits	17.84
iv)	Lead naphthenate (24%)	0.39
	Cobalt naphthenate (6%)	0.30
		100.00

PROCESS

Disperse pigment in alkyd (A) overnight on a ball mill. Add ii). Dissolve Resen 95 – 105 in solvent iii) and add to alkyd mix. Add iv).

ALKYD RESIN FOR LITHOGRAPHIC INKS

i)	Linseed oil (alkali refined)	40.50
	Neopentyl glycol	4.80
	Litharge	0.02
ii)	Imprez 125C	54.70
		100.00

PROCESS

Charge i) and heat to 260°C. Hold for 1 hour at 260°C. Add ii) and allow to cool to 250°C. Hold at 250°C for about five hours until the acid value is less than 10mg KOH/g.

Imprez 125C is a carboxylated aliphatic resin, which is claimed (4) to function as a hard resin and is the acid modifying component of an alkyd.

The 40% oil length alkyd is then blended with a variety (or mixture) of oils, which can include linseed, soya, tung and oiticica, to the required oil length (typical range 50−60%). With linseed and soya, oil lengths of 70 and 80% are possible. All of these systems are reducible in white spirit or other aliphatic hydrocarbon solvents. Drying is by a combination of physical and chemical effects, thus forced drying is necessary.

EMULSIFIABLE VERSION OF ALKYD RESIN

By including PEG 400 as the hydroxylic component a readily emulsifiable version of alkyd (4) can be obtained.

i)	Linseed oil (alkali refined)	31.20
	PEG 400	17.00
	Litharge	0.02
ii)	Imprez 125C	51.80
		100.00

PROCESS

Process as for the above alkyd. However, it should be noted that the oil length has been reduced from the alkyd by 10% (i.e. from 40 to 30%).

**GENERAL PURPOSE PRIMER AND
ANTI-CORROSIVE PAINT MEDIA**

Formulation of Oleoresinous Vehicle	
i) Petrosin 120	48.0
Boiled linseed oil	20.0
ii) Mineral spirit	32.0
	100.0

PROCESS

Charge i) and heat to 230−240°C. Hold at 240°C for 10 mins and charge ii). Cool. For use as a varnish add driers, etc.

**GREY PRIMER
(For a grey primer the following
formulation can be used)**

Titanium di-oxide	15.0
Carbon black	0.1
Calcium carbonate	17.0
Aluminium stearate	0.2
Oleoresinous vehicle from above	64.0
Driers	0.5
Methyl ethyl ketoxime	0.2
Mineral spirit	3.0
	100.0

**ESTERIFIED PHENOLIC RESIN
(SOFT) LINSEED OIL VARNISH**

Phenolic resin	33.4
Linseed stand oil	66.4
	100.0

PROCESS

The rosin phenolic esterified resin with a melting point of ca 25°C is added to a medium viscosity linseed stand oil at 150−200°C. Reaction is complete when a clear dispersion is obtained in white spirit.

ESTERIFIED PHENOLIC RESIN (HARD) LINSEED OIL VARNISH

i)	Phenolic resin	33.4
	Linseed stand oil	33.3
ii)	Linseed stand oil	33.3
		100.0

PROCESS

The harder rosin phenolic esterified resin (melting point ca 150°C) is added to the oil at 250°C and the temperature is gradually increased to 280°C. When a cold bead clears the reaction is complete and ii) added. China wood oil could be used instead of linseed stand oil.

NOVALAC TUNG OIL WATER RESISTANT VARNISH

i)	Novalac phenolic (p. phenyl)	24.95
	Tung oil	24.95
ii)	Xylene	45.74
	Butanol	4.15
	Magnesium naphthenate	0.21
		100.00

PROCESS

Charge i) and heat to 150°C over 30 mins. Hold at 150°C for 60 mins and then heat to 230°C over 20 mins. Hold at 230°C for 50 mins. Cool to 180°C over 15 mins and add ii). Drying time is 30−60 minutes.

RESOLE TUNG OIL
WATER RESISTANT PRIMER MEDIUM

i)	Tung oil	35.67
ii)	pt butyl resole	9.91
iii)	Linseed stand oil (30 poise)	3.96
iv)	Mineral spirit	37.65
	Xylene	12.55
	Cobalt naphthenate	0.26
		100.00

PROCESS

Charge i) and heat to 170°C over 20 mins. Charge ii) and allow to cool back to 140°C over 20 mins by which time the reactive phenolic resin should have dissolved. Heat to 260°C over 45 mins and add iii). Cool to 230°C over 5 mins and then to 170°C over 20 mins and add iv).

PHENOLIC RESIN LINSEED OIL VARNISH FOR
LETTERPRESS OR LITHOGRAPHIC APPLICATIONS

Kelrez 42404	45 – 55
Linseed oil (alkali refined)	20 – 30
Petroleum solvent	20 – 30
	100

PROCESS

Kelrez 42404 is a rosin modified phenolic resin available from Kelrez Corn Products (10). An alternative approach is to incorporate a rosin ester into the oleoresinous media.

Kelrez 42376	25 – 35
Kelrez 42404	5 – 15
Linseed oil (alkali refined)	25 – 35
Petroleum solvent	25 – 35
	100

HYDROCARBON LINSEED OIL VARNISHES FOR PRINTING INKS

Neochem 140 **or**	
Necires LX-685	66.6
Linseed oil	33.4
	100.0

PROCESS

Neochem 140 is an alkyd aromatic hydrocarbon resin and Necires LX-685 is a cyclo aliphatic, aromatic hydrocarbon resin. Both are available from Neville Cindu Chemie (Holland). The viscosity of the oleoresinous varnish depends upon the cooking time.

Necires gives a higher viscosity for the same length of cooking as Neochem. This is to be expected, because the iodine value of the former is three times that of Neochem. This demonstrates that some degree of reaction has occurred during cooking. These media find application in offset, heat and cold set inks.

ROSIN MALEIC GLYCEROL/OIL MEDIA

Mitchanol 56	50
Oil	50
	100

PROCESS

Heat Mitchanol 56 (Ex-Warwick Chemicals) which is a high viscosity rosin modified maleic resin esterified with glycerol and oil to 250°C until a clear cold bead is formed. Additional oil if required is then added and the temperature maintained at 250°C. When the desired viscosity is reached, cool and thin in mineral spirit. Any of the drying oils can be used.

AN ALTERNATIVE METHOD OF MODIFYING OILS WITH RESINS

Monoglyceride Process

Many of the varnishes made with natural resins have relatively high acid values which cause problems when used with basic pigments. The gelation which can sometimes result is called 'livering'. Using the ester of the resin to form an oleoresinous vehicle reduces this acidity.

An alternative approach is the monoglyceride process. An example would be heating fused copal and stand oil 50/50 with sufficient linseed monoglycerides to neutralise the acid. A lime catalyst is used. The reaction is held at 270−280°C until the desired acid value is reached. The mixture is then oiled down and let down in solvent in the normal manner. Care must be exercised in selecting the grade of copal to be used. Slack melts can be esterified in this manner.

Pentaerythritol can also be used to reduce acidity when stand oil is present but blooming can be a problem.

REFERENCES

1. Hydrocarbon and Modified Hydrocarbon Synthetic Resins. Published by Hercules.
2. Hercules Resins for Printing Inks. Published by Hercules.
3. Petrosin booklet. Published by Mitsui Petrochemical Industries Limited.
4. Imprez Technical Bulletin. Published by ICI Petro Chem. Division, Middlesbrough, U.K.
5. 'Britrez — A Range of Versatile Hydrocarbon Resins'. Published by British Steel Corporation, U.K.
6. Coumarone — Indene Resins, etc. Published by UFT part of the Nutgerswerke AG−Tar Chem. Division.
7. Resins for Paint Lacquer and Printing Ink. Published by Laporte Industries Limited. Organic and Pigment Division.
8. Technical Bullctins, Reichhold Chemical Inc. New York.
9. C.D. Smith, & K. Wise, J. Paint Tech. *41* 338, (1969).
10. Kelrez Corn Products.
11. Paint Technology Manual Part 2. Published by OCCA in 1961.
12. D.H. Solomon 'The Chemistry of Film Formers'. Published by R. Krieger, New York 1977.
13. Surface Coatings. Vol. 1 p. 102. Published by OCCA 1983.
14. 'Printing Ink Technology' by E.A. Apps. Published 1961 by Leonard Hill Books.
15. Amberol Synthetic Resins for Coatings. Published by Rohm & Haas.
16. Drying Oil Technology by M.B. Mills. Published by Pergamon Press (London) 1952.

Chapter IV

Alkyd Resins

Chapter IV

Alkyd Resins

Alkyd resins have found applications in surface coating systems for the last forty years. At present they hold a majority share of the world market for non-aqueous binders. The major reasons for this popularity are:

COST

Alkyds are relatively inexpensive in terms of raw material and manufacturing costs and since they are also readily soluble in the less expensive organic solvents, they provide the surface coating formulator with a relatively low cost pigment vehicle.

VERSATILITY

Whilst other polymer types may offer improved properties in specific areas, alkyd resins have the widest spectrum of acceptable properties in terms of surface coating usage. They are easy to pigment and are compatible with most substances used in surface coatings, e.g. nitrocellulose, amino resins, phenolics and polyurethanes. In addition they are tolerant of most substrates and are easily modified for specialist applications. The alkyd is the 'work horse' of the coatings industry, finding applications in:

Printing inks

Decorative paints

Air-drying and stoving industrial paints

Metal decorating coatings, also, a variety of other areas, including water soluble and electro-deposition systems.

Since they are highly tolerant to misuse, they are also frequently found in areas of application where the use of more suitable polymer systems would be cost prohibitive.

Alkyd resins are essentially short branched polyester chains, formed by the polycondensation of a di-basic acid and a polyhydric alcohol, in the presence of a glyceride oil or oil derived fatty acid. The presence of the oil confers good pigment wetting properties and, when unsaturated, allows coherent films to be formed on cure. The polyester chain confers hardness and durability to the film and improved drying speed.

The presence of the oil also adversely affects colour and gloss retention, whilst the nature of the ester linked chain means that the film can be attacked by strong alkalis and acids.

This chapter outlines the basic chemistry of alkyd and modified alkyd resin manufacture and attempts to relate properties and uses to chemical structure and resin type.

Starting formulations for both resins and end use products are given as illustrations of the versatility of alkyd resins.

TECHNIQUES USED IN ALKYD RESIN MANUFACTURE

Glyceride oils (with the exception of castor oil, which contains a hydroxyl group on the fatty acid), are not readily reactable with the polyol and acid components of an alkyd. Although during the polycondensation reaction some of the oil is split into its component fatty acids and glycerol (and is thus randomised along the polyester chain), the acid reacts preferentially with polyol, to give a heterogeneous mixture of polyester and free oil. This can be overcome by:

Alcoholysis of the Oil

This is a technique by which the oil is pre-reacted with a polyol, to give a product capable of taking part in the polycondensation reaction. During alcoholysis ester interchange is brought about between the polyol and oil. The tri-glyceride oil is converted into reactive monoglyceride.

The idealised equation below shows the reaction between an oil and glycerol. A similar reaction occurs with other polyols.

$$
2\ \begin{array}{l} CH_2 -\!-\ OH \\ | \\ CH - OH \\ | \\ CH_2 -\!-\ OH \end{array}
\ +\
\begin{array}{l} CH_2 -\!-\ OOCR \\ | \\ CH - OOCR \\ | \\ CH_2 - OOCR \end{array}
\ =\
3\ \begin{array}{l} CH_2 - OOCR \\ | \\ CH - OH \\ | \\ CH_2 - OH \end{array}
$$

glycerol   triglyceride (oil)   monoglyceride

In practice, complete monoglyceride formation is not possible and is unnecessary, di-glycerides and small amounts of tri-glyceride and glycerine can be accommodated, provided that randomisation can occur during the polycondensation stage.

The typical composition of a commercial linseed, glycerol partial ester is 51% monoglycerides, 40% di-glycerides, 4% tri-glycerides and 5% glycerine.

Alcoholysis requires temperatures of 240°C−260°C and ester interchange is greatly speeded up by the use of basic catalysts. Originally litharge was commonly used as an alcoholysis catalyst, but this has now been replaced by lead acetate and more recently, by the even more effective lithium hydroxide (sodium hydroxide is also used commercially). The level of catalyst normally used is 0.03−0.04% w/w on the oil charge for lead acetate and 0.01−0.03% for lithium hydroxide. In practice lead salts tend to give a hazy suspension, particularly with linseed oil. Although this alcoholysis haze can be removed by efficient filtration, lithium and sodium are the preferred catalysts.

Alcoholysis is normally carried out, with the reactor under a slight positive inert gas pressure to prevent ingress of air, which would otherwise result in discolouration of the product. The progress of the ester interchange reaction is monitored by measuring the tolerance of the reaction mixture towards alcohol. Initially the mixture of oil and polyol has a low tolerance, but as the level of tri-glyceride decreases, so the alcohol tolerance of the reaction mixture increases. It is important that the tolerance required for a particular system by accurately determined beforehand, since short oil length alkyds are far less able to accommodate unconverted tri-glyceride than long oil length alkyds. Incomplete alcoholysis will result in the formation of insoluble esters, on addition of the acid.

Acidolysis of the Oil

This is the technique whereby the oil is pre-reacted with fatty acid:

$$
\begin{array}{l}
CH_2 - OOCR' \\
| \\
CH \ - OOCR'' \quad + \\
| \\
CH_2 - OOCR'''
\end{array}
\quad
\begin{array}{c}
COOH \\
\bigcirc \\
COOH
\end{array}
\quad \rightleftarrows \quad
\begin{array}{l}
\qquad\qquad COOH \\
\qquad \bigcirc \\
CH_2\,OOC \\
| \\
CH\,OOCR'' + R'\,COOH \\
| \\
CH_2\,OOCR'''
\end{array}
$$

Temperatures of over 260°C are required for acidolysis and although tin catalysts may be used to speed up the reaction rate, the process takes longer than that for alcoholysis. As a result of the higher temperature and longer process time there is more risk of colour deterioration and polymerisation of the oil, than with the alcoholysis process.

Acidolysis is normally used, only where there are problems associated with the solubility or reactivity of the di-carboxylic acid, which may be overcome by pre-reaction with the oil.

The Fatty Acid Process

As the name suggests this technique utilises fatty acids derived from the oils, rather than the oils themselves.

The carboxylic group on the fatty acid is able to react directly with the polyol. Thus the fatty acid, polyol and di-acid can be charged together to the reactor and polycondensation brought about without the need for alcoholysis or acidolysis.

Although fatty acids are more expensive than oils, the shorter process times conferred by the fatty acid process are attractive, when the economics of the various process options are considered.

A frequently used compromise is to employ a mixture of fatty acid and oil. Providing that the ratio of fatty acid to oil is restricted to one that yields a homogenous reaction mixture and the polycondensation reaction is not too rapid to prevent randomisation of the oil, then no alcoholysis or acidolysis stage is required.

Advantages and Disadvantages of the use of the Fatty Acids as opposed to Oils

ADVANTAGES

i) Lighter coloured products.

ii) Shorter process time.

iii) Synthetic fatty acids and fatty acids (such as tall oil), not normally obtained as triglyceride oils, can be included in the alkyd resin formulation.

iv) Freedom of choice of polyol. When an oil is used in an alkyd resin, part of the overall polyol content is provided by the glycerol from the oil. Where fatty acids are used, glycerol may be ommitted entirely from the formulation if desired.

DISADVANTAGES

i) More expensive raw material costs.

ii) Separation of the more saturated components of the fatty acids can occur, on storage at low temperature. Therefore heated storage tanks have to be provided for the fatty acids.

POLYCONDENSATION

The polycondensation reaction consists of a series of simple esterification reactions.

A hydroxyl group reacts with a carboxyl group to form an ester link with the elimination of a molecule of water.

$$\sim\sim\sim R - OH \quad + \quad HOOC - R^1 \sim\sim\sim$$

$$\downarrow$$

$$\sim\sim\sim R - OOC - R^1 \sim\sim\sim$$

$$+$$

$$H_2O$$

An idealised alkyd structure based on glycerol is shown below:

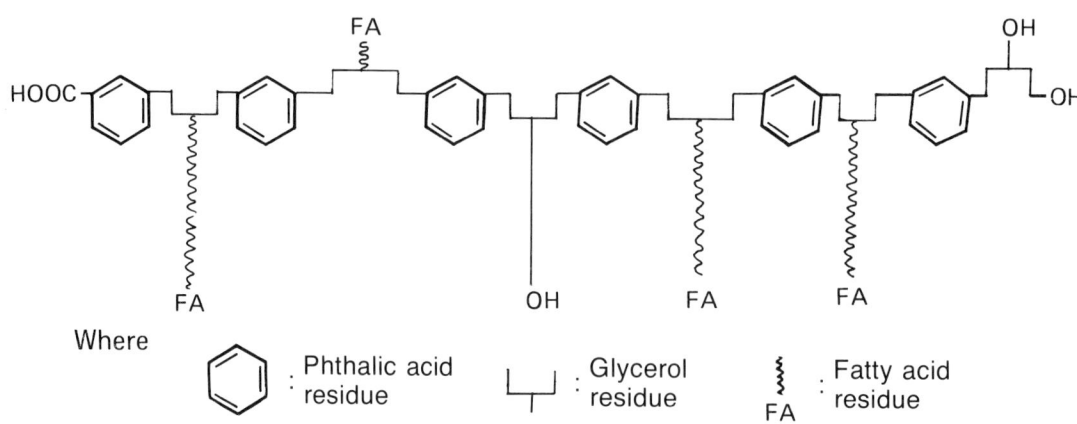

Where
☐ : Phthalic acid residue
⊥ : Glycerol residue
ξ FA : Fatty acid residue

Two techniques are used commercially to manufacture alkyd resins:

i) Fusion process

ii) Solvent process.

The Fusion Process

The reactants (after alcoholysis, if an oil is used) are heated together to a temperature of between 180°C and 260°C. The exact temperature is determined by the rate of reaction desired. Below ca 180°C esterification is very slow, while above 260°C, the rapid reaction rate, coupled with the tendency of unsaturated fatty acids to heat polymerise, can prevent the necessary degree of process control being exercised.

A flow of inert gas is maintained through the reactor, to prevent ingress of air and to remove water of the reaction.

There is a considerable loss of volatile reactants during a fusion process, particularly at higher temperatures.

Allowance has to be made for these losses in the initial reactor charge.

Generally the fusion process is used only for longer oil length formulations, where unanticipated losses will have a less serious effect on the predicted functionality and hence the practical gel point of the resin.

The water eliminated from the reaction, together with volatile reactants, exits from the reactor via fume trunking, fitted with a fume-scrubbing system, before being exhausted to atmosphere. Efficient removal of the water of reaction is essential to the progress of the polycondensation and can be aided by application of negative pressure (vacuum) to the fume extraction system.

The Solvent Process

The reactants are heated, together with a solvent (usually xylene), which aids the removal of the water of reaction. The solvent and water are distilled from the reactor, condensed and passed into a separator. Here the water is removed to waste and the solvent recycled back into the reactor.

This technique is the preferred mode of manufacture, when precise specification and performance requirements must be met by the alkyd. The use of solvent distillation to aid water removal, means that short process times can be achieved without recourse to excessively high temperature (typically solvent process temperatures are in the range 200–240°C). The solvent vapour blanket, effectively prevents ingress of air, thus enabling low colour products to be produced with a minimum of inert gas usage.

Most of the volatile reactants, lost from the reactor, are returned with the solvent. Thus a more uniform resin composition, with a narrower molecular weight distribution can be obtained. This, in turn, means that a more accurate prediction of the practical gel point can be made, allowing the resin formulator to produce higher molecular weight resins, with a corresponding improvement in drying and film performance properties.

Polycondensation is continued, until the desired 'degree of esterification' (that is the required molecular weight at the specified residual acid and hydroxyl value) has been achieved. When this point is reached the alkyd is cooled to below 180°C, to prevent further reaction and diluted with solvent, to the required solids or non-volatile content.

At the start of polycondensation, the reactants exist as separate molecules. These combine to form small units, based on two or three of the original molecules and as the reaction proceeds, so larger and larger units are formed. This means that initially, the increase in resin viscosity per unit decrease in acid value is small.

As the degree of esterification increases however, and larger units combine together, so the viscosity increase per unit decrease in acid value, becomes larger.

The concentration of OH and COOH groups at the start of the reaction is high.

There is a high probability that these groups will come close enough together to allow reaction. Thus the acid value falls rapidly with time. However, as the reaction proceeds, the concentration of reactive species is decreased and those species that are present are now incorporated in large molecular units.

Steric hindrance and restricted mobility factors, also play a part in reducing the availability of the reactive species. Thus the acid value falls more slowly with time, whilst the viscosity increases more rapidly with time.

This is illustrated below in graphical form:

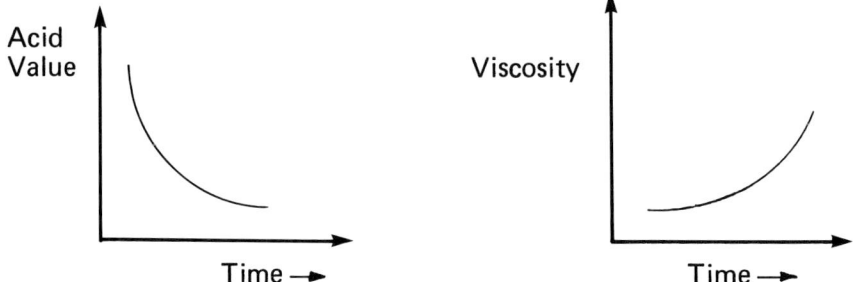

In practice, the course of the reaction is followed by monitoring the increase in viscosity of the resin (which reflects the increase in molecular weight) and by the determination of acid value. The hydroxyl value is not normally measured during processing, since, (assuming no loss of reactants), the hydroxyl value is fixed for any given acid value by the hydroxyl excess employed in the resin formulation.

A plot of acid value versus viscosity for a particular resin formulation is used as a standard against which the determined resin constants are compared. To be of any value, the standard plot must have been under the same conditions (e.g. process temperature, reactor charge etc.), as the resin with which it is compared. An example of such a plot follows:

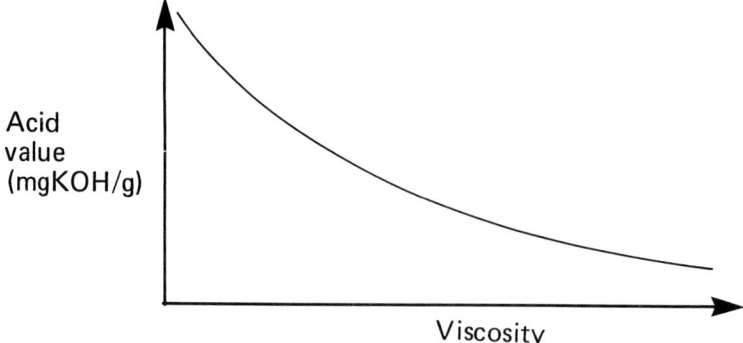

The use of a standard plot, allows any abnormal viscosity/acid value relationship to be spotted immediately and an in-process correction made, to adjust the functionality of the resin and to bring about a return to the standard relationship.

Where the polycondensation follows a course above the standard curve, the system is overfunctional and corrective additions, designed to decrease the functionality of the resin, are required.

Where the course of the reaction is displaced below the standard curve, the system is underfunctional. Additions designed to increase the functionality of the resin are required.

Since any long term departure from intended functionality will result in a product with a non-standard molecular weight profile, it is important that all corrective additions be made as soon as possible after the anomoly has been discovered.

Alkyd Resin Functionality

To produce a polymer by esterification, it is necessary that both the alcohol and acid have at least two reactable groups each.

Consider the diol $HO-R^1-OH$ and the di-acid $HOOC-R''-COOH$. Each molecule has two reactable groups, therefore each molecule has a potential functionality of 2.

If we bring about reaction between equimolar quantities of diol and di-acid and continue the esterification until the maximum reaction is obtained, then we would produce a linear polyester chain containing the total number of molecules present.

In such a system each molecular species is able to achieve its maximum functionality of 2 and the overall functionality of the system is also equal to 2.

Consider the same reaction using 1 mole of di-acid and 2 moles of diol.

$$2HO-R-OH \quad + \quad HOOC-R''-COOH$$

$$\downarrow$$

$$2H_2O \quad + \quad HO-R'-OOC-R''-COO-R'-OH$$

In this case, the polyol molecules can achieve only an average actual functionality of 1, since there are only enough acid groups to react with 2 out of every 4 OH groups.

Thus, the actual functionality of a species, is related to the mole ratio of the reactants present and actual and potential functionality is not necessarily the same.

The actual functionality of the polyol, in the above system, can be calculated using the following:

$$F \text{ (actual)} = \frac{F \text{ (Potential)}}{1 + \dfrac{E}{100}}$$

Where F (potential) is the maximum functionality, if all the groups react and E is the % of the OH groups in excess of the available COOH groups.

Thus from the above:

$$F \text{ (actual)} = \frac{2}{1 + 0.5} = 1.33$$

% OH excess is calculated as:

$$\frac{OH \text{ equivalents in excess} \times 100}{OH \text{ equivalents}}$$

Where OH equivalents in excess are:

$$(\text{Total of OH equivalents}) - (\text{total } CO_2H \text{ equivalents})$$

The overall average functionality of a system may be calculated from:

$$F \text{ (average)} = \frac{\text{Total No. of reactable equivalents present}}{\text{Total No. of moles present at start of reaction}}$$

For an OH excess system this becomes:

$$F \text{ (average)} = \frac{2 \times COOH \text{ equivalents present}}{\text{Total moles present at start of reaction}}$$

And for our examples above:

$$F \text{ (average)} = \frac{4n}{2n} = 2 \text{ for the 1st case}$$

and

$$F \text{ (average)} = \frac{4}{3} = 1.33 \text{ for the 2nd case}$$

The use of either an excess of one of the reactants, or the addition of a monofunctional reactant to act as a 'chain stopper', assumes great importance in the more functional systems encountered in alkyd resin manufacture.

In a system where a tri-functional or higher species is present, cross-linking and gelation are possible.

Take for example a system employing tri-methylol propane and phthalic anhydride. If we arrange for equivalent concentrations of OH and COOH (that is a mole ratio of 3 moles phthalic to 2 moles of tri-methylol propane) each individual molecule can achieve its maximum potential functionality.

Trimethylol Propane   Phthalic Anhydride

The functionality of such a system would be 2.4 and it can be expected that chain branching and cross-linking between reactive groups on neighbouring chains will occur, leading to the formation of an insoluble high molecular weight solid, i.e. gelation will occur.

$$F \text{ (average)} = \frac{\overset{\text{TMP}}{[6]} + \overset{\text{PA}}{(6)}}{5} = \frac{12}{5} = 2.4$$

It is necessary to formulate systems with a functionality as near 2 as possible, in order to produce a resin capable of satisfactory performance in a surface coating system and retain solubility in inexpensive organic solvents. Where reactants with a potential functionality greater than 2 are present (e.g. glycerol, pentaerythritol, tri-methylol propane, etc.) it is necessary to restrict the actual functionality of the system, by employing an equivalent excess of one of the high functionality species, or by inclusion of monofunctional reactants (e.g. fatty acid, benzoic acid, etc.) or by employing both methods.

The use of a monofunctional material, as a 'chain stopper', can be illustrated by replacing 1 mole of phthalic anhydride with 2 moles of fatty acid in the formulation discussed above.

The actual functionality of the tri-methylol propane is still 3, as before, but the functionality of the overall system is now only 2.

$$F \text{ (average)} = \frac{TMP [6] + PA (4) + FA (2)}{6} = \frac{12}{6} = 2$$

However, the above depicts an idealised structure. In practice, some tri-methylol propane molecules would be expected to react with more than one molecule of fatty acid, thus preventing linear chain growth for that unit; while other molecules will not have reacted with any fatty acid, thus being free to branch and cross-link as before.

So that in systems where species of more than one functionality are present, the calculated functionality of the system is only an average figure and only depicts the probable reaction course for the majority of polymer chains. Small amounts of highly cross-linked material and very low molecular weight material will also be present.

When manufacturing commercial alkyd resins, it is essential to obtain information on the degree of reaction that can be obtained with a particular formulation, before gelation is likely to occur.

A simple treatment, based on the work of Carothers and others, results in the equation:

$$P \text{ (gel)} = \frac{2}{f \text{ (average)}}$$

Where P(gel) is the degree of reaction at gelation and $P = 1$ at 100% reaction. Thus an alkyd system with an f average of 2.04 would be expected to gel at:

$$P \text{ (gel)} = \frac{2}{2.04} = 0.98$$

i.e. 98% of complete reaction

The theoretical acid value at start of reaction can be converted into acid value at gelation as follows:

If acid value at start of reaction = 200 mgKOH/g

then acid value at gelation will be

$$200 - (0.98 \times 200) = 4 \text{ mgKOH/g}$$

Carothers definition of gel point is based on the number average molecular weight Mn. In a polydispersed system, such as an alkyd resin, the weight average molecular weight (Mw) will give a much more accurate description of the system in terms of branching and cross-linking.

Thus a more accurate description of the gel point of an alkyd resin, can be obtained from the following equation, quoted by Solomon and derived originally from the work of Flory and others. It is based on the concept that the weight average molecular weight, is much larger than the number average molecular weight, at the onset of gelation.

$$P(gel) = \sqrt{\frac{\Sigma}{2(1-\lambda)}}$$

where Σ = $\dfrac{\text{Total equivalents of OH present}}{\text{Total equivalents of COOH present}}$

and λ = $\dfrac{\text{equivalents of COOH from monofunctional acid}}{\text{Total equivalents of COOH present}}$

Although the above equation is more accurate than the Carothers derived equation, in predicting P(gel) for an alkyd system, neither equation gives absolutely reliable results.

The reasons for the lack of precise practical agreement with the P(gel) derived from the above can be summarised as follows:

1. Volatile material may be lost from the reactor during processing and effectively alter the average functionality of the system.

2. The calculated functionality is only an average figure. Reactions do occur in practice, which are not allowed for in the theoretical calculation. These effectively change the functionality of the system and include the following:

i) Ring closure between two species, effectively preventing chain propagation:

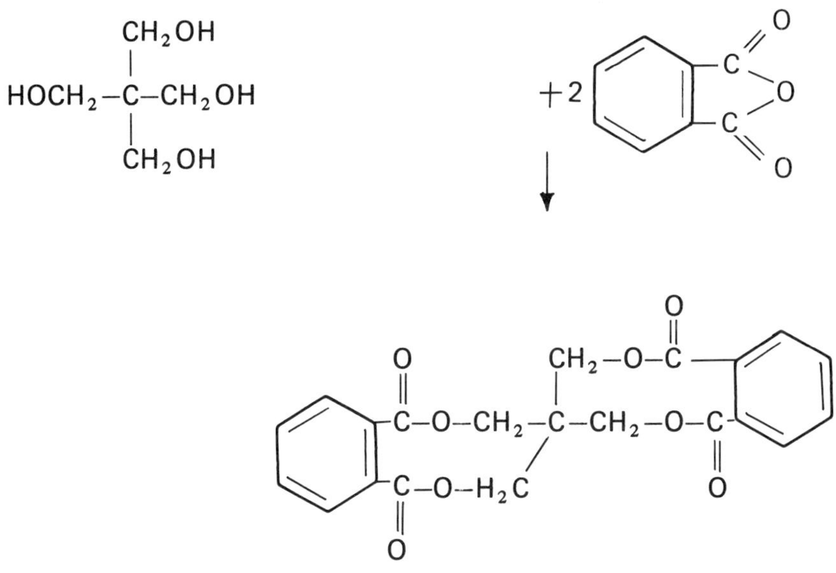

$$CH_2OH$$
$$|$$
$$HOCH_2-C-CH_2OH$$
$$|$$
$$CH_2OH$$

$+2$ phthalic anhydride

↓

pentaerythritol tetra-phthalate ester (cyclic structure)

ii) Polymerisation between molecules of one species:

$$2 \begin{array}{ccc} CH_2 & - CH & - CH_2 \\ | & | & | \\ OH & OH & OH \end{array} \rightarrow \begin{array}{cccccc} CH_2 - CH - CH_2 - O - CH_2 - CH - CH_2 \\ | \qquad | \qquad\qquad\qquad\qquad | \qquad | \\ OH \quad OH \qquad\qquad\qquad\quad OH \quad OH \end{array}$$

iii) Reactions between two species that do not involve the OH and COOH groups:

$$\sim\!\!\sim\!\!\sim CH_2-CH{=}CH-CH{=}CH-CH_2 \sim\!\!\sim\!\!\sim$$

$$+$$

$$\downarrow$$

3. All groups are assigned equal reactivity. This is not the case.

   i) Glycerol has 2 primary hydroxyl and 1 secondary hydroxyl groups. The secondary hydroxyl will react more slowly than the primaries, yet a potential functionality of 3 is assigned to the glycerol molecule. In practice a more accurate figure would lie between 2 and 3 but nearer 3.

   ii) As the polymer chain lengths increase, the reactivity of some groups may become restricted by steric hindrance. This affects the actual functionality of the system. When sterically hindered groups are OH groups, the effective OH excess is reduced and P(gel) is also reduced.

   When the sterically hindered groups are COOH groups, the effective OH excess is increased and P(gel) is also increased.

4. The calculated gel point is taken as the onset of branching. In practice, significant amounts of three dimensional polymer networks are present at gelation and branching does occur well before gelation. This does not mean that the calculated gel point should be ignored when formulating an alkyd resin, rather it means that the information should be used as a guide to aid experimentation at laboratory scale — in preparation for factory scale manufacture.

THE CLASSIFICATION OF ALKYD RESINS

Alkyd resins are characterised in terms of oil length and oil type.

Oil Length

This is the amount of oil (or fatty acid expressed as tri-glyceride oil) present as a percentage of the total non-volatile content. This term is left over from the days when surface coatings were little more than pigments dispersed in modified tri-glyceride oils. Although it does little to describe the complex polymer system which is the modern alkyd, it is universally used throughout the surface coating industry and as such is a convenient method of classification.

Using this system, alkyds can be classified as:

Long oil — oil content greater than 55% w/w

Medium oil — oil content 45−55% w/w

Short oil — oil content less than 45% w/w.

Oil Type

Alkyds may be further classified by oil type as shown below:

OXIDISING ALKYDS

An oxidising alkyd contains drying or semi-drying (unsaturated) oils or fatty acids, and is able to film form by oxidation (air-dry). This type of alkyd usually has an oil length in excess of 45%.

NON-OXIDISING ALKYDS

A non-oxidising alkyd contains non-drying (saturated) oils or fatty acids and is not capable of coherent film formation by oxidation. They are reacted with other polymer types to produce 'cured' films, e.g. amino resin/alkyd resin blends. Non-oxidising alkyds usually have oil lengths below 45%.

Not all oxidising alkyds are used in air-drying systems and not all non-drying alkyds are 'cured'. The following table correlates main areas of use with alkyd type and type of cure employed.

OXIDISING ALKYDS

Oil Length	Typical Oil (or fatty acid)	Cure	Use
45% to 60%	Linseed, Soya, Safflower Dehydrated Castor, Sunflower, Tall Oil (fatty acid) and Tung Oil*	Air drying	Decorative paints and printing inks
Less than 45%		Stoving and co-cure systems with amino or phenolic resins etc.	Industrial coatings
	*Tung Oil is rarely used on its own, but in combination with other oils.		

NON-OXIDISING ALKYDS

Oil Length	Typical Oil (or fatty acid)	Cure	Use
40% to 60%	Coconut, Castor and short chain length fatty acids. Commercially available synthetic blends of non-drying fatty acids are frequently used.	Non-curing	Plasticiser for other resin systems
Less than 45%		Co-cured with amino or phenolic resins	Reactive plasticiser in metal decorating and industrial coatings.

The Effect of Oil Length and Oil Type on the properties of an Alkyd Resin

The characteristic of an alkyd resin and in particular the film properties, are related to the oil type and oil length. The longer the oil length, the more the alkyd reflects the properties of the oil and the shorter the oil length, the more the properties of the polyester chain predominate. This is illustrated below in a diagram taken from available literature.

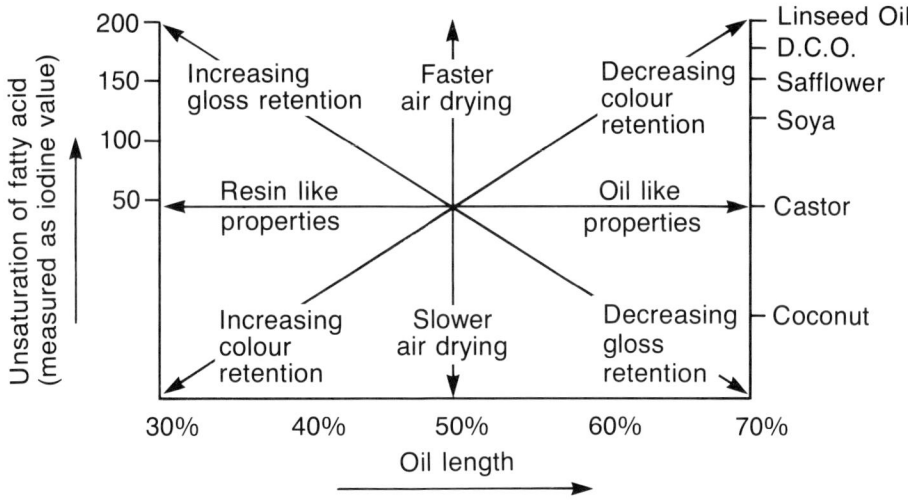

When selecting an alkyd for a particular application, it is necessary to specify both oil length and oil type, to arrive at a balance of the properties derived from the oil (e.g. solubility in cheap solvents, ease of pigmentation) and those derived from the polyester (e.g. improved colour, durability).

TYPES OF ALKYD

Air Drying Alkyds

The unsaturation in the fatty acid chains, is largely unaffected during alkyd resin manufacture and thus the mechanism of film formation (drying) is essentially that of the tri-glyceride oils.

Although the presence of hydroxyl groups on the alkyd chain should, in theory at least, retard hydroperoxide breakdown and thus retard the drying (see chapter on glyceride oils), in practice the presence of residual carboxyl groups on the alkyd, tend to promote hydroperoxide breakdown and no noticeable overall effect is observed, unless the residual hydroxyl value is in excess of 60−70 mgKOH/g and the residual acid value is below 5 mgKOH/g.

The molecular weight of an alkyd is considerably higher than that of an oil, so that fewer cross links are required, before a coherent film is formed. It follows from this that resins dry very much more rapidly than the corresponding tri-glyceride oil.

This allows the alkyd resin formulator to employ semi-drying oils such as soya bean, tall oil fatty acid, sunflower oil, etc. in an alkyd and yet have rapid drying capability.

As illustrated in the table in Section 4 the semi-drying oils have less tendency to discolour on drying, than the highly conjugated drying oils. Thus by accepting a small decrease in drying efficiency, a considerable colour improvement can be obtained. Oil lengths of 50−60% are normally used for this type of application, but it must be remembered that, although the higher oil lengths bring improved drying rates, they also reduce gloss and colour retention as well as durability.

Force Drying Alkyds

The oxidation of fatty acids, is greatly accelerated by increased temperature. Alkyds based on unsaturated fatty acids and oils, can be cured rapidly at temperatures in the range $120°-160°C$. Under these conditions a high percentage of the cross links are of the $C-C$ bond type (see chapter on glyceride oils) and the resultant films are more chemically resistant and durable than those formed with a high percentage of $C-O-O-C$ cross links (as with air drying systems).

The faster rate of cure, means that shorter oil lengths can be used and the consequent increase in polyester content, results in improved durability, gloss retention and colour. Oil lengths of $40-50\%$ are normally employed for this type of resin and the oils are usually selected on the basis of a compromise between colour retention and drying properties.

Plasticising Alkyds

Alkyd resins can be used as non-reactive diluents, in combination with other resin types. Their function is to act as a plasticiser, giving the combined resin system increased flexibility, adhesion and gloss.

Plasticising alkyds are used to good effect, in combination with amino modified finishes and nitrocellulose systems. They are particularly effective in nitrocellulose based vehicle refinishing and wood finishing systems. Nitrocellulose films normally have poor adhesion, flexibility and gloss, but in combination with a suitable plasticising alkyd, high build coatings, with good adhesion, flexibility and high gloss can be obtained.

Unlike castor oil which is commonly used with nitrocellulose, the alkyd is a true solvent for nitrocellulose and will not exude from the film, as other non-polymeric plasticisers tend to do.

Since these resins are not intended to film form on their own, they are usually formulated on non-drying oils or fatty acids. Most commonly used oils are castor oil, and castor oil derived fatty acids, but the more expensive coconut oil and coconut oil fatty acid substitute are often used, in addition to short chain acids like pelargonic acid.

Plasticising alkyds may also be used as mill bases, where the good pigment wetting properties of the alkyd, coupled with their excellent compatibility with other resin types, can be exploited, to allow the pigmentation of resins which would otherwise prove difficult to pigment.

Curing Alkyds

Amino resins (reactive melamine-formaldehyde and urea-formaldehyde resins) and phenolic resins (phenol-formaldehyde resoles) can be used in combination with alkyd resins, to form a co-cured surface coating system. The amino resin/alkyd resin combination results in films with improved hardness, exterior durability, alkali resistance and cure time. Cure involves the reaction of the residual hydroxy groups of the alkyd resins with reactive groups on the amino or phenolic resin.

Residual carboxylic acid groups on the alkyd, can also undergo these reactions but more slowly.

The presence of carboxylic acid groups on the alkyd, tend to catalyse reactions involving the hydroxyl groups (see below). The ratio of amino resin to alkyd is normally 1:3 and the alkyd functions as a reactive plasticiser.

In general, urea and melamine-formaldehyde resins are used with alkyd for top-coats and phenolics are used with alkyd for undercoats and primers, where the darker colours they produce, can be offset against their improved chemical resistance.

Cure Reactions with Amino and Phenolic Resins

a) The reaction of OH groups on the alkyd chains, with methylol groups in amino or phenolic resins can be shown as:

$$\boxed{R} - CH_2 - OH \ + \ \sim\sim\sim\boxed{A}\sim\sim\sim$$
$$\qquad\qquad\qquad\qquad\qquad |$$
$$\qquad\qquad\qquad\qquad\quad OH$$

$$\Big\downarrow H+$$

$$\sim\sim\sim\boxed{A}\sim\sim\sim$$
$$\qquad\quad | \qquad\qquad + \quad H_2O$$
$$\boxed{R} - CH_2 - O$$

Where $\boxed{R}$ is an Amino or phenolic resin residue and $\boxed{A}$ represents an alkyd resin chain.

b) The reaction of OH groups with etherified methylol groups can be shown as:

$$\boxed{R} - CH_2OR^1 \ + \ \sim\sim\sim\boxed{A}\sim\sim\sim$$
$$\qquad\qquad\qquad\qquad\qquad\qquad |$$
$$\qquad\qquad\qquad\qquad\qquad\quad OH$$

$$\Big\downarrow H+$$

$$\sim\sim\sim\boxed{A}\sim\sim\sim$$
$$\qquad\quad | \qquad\qquad + \quad R^1 OH$$
$$\boxed{R} - CH_2 - O$$

Where R^1 is a short chain alcohol.

148

c) The reaction of COOH groups with etherified methylol groups can be illustrated as:

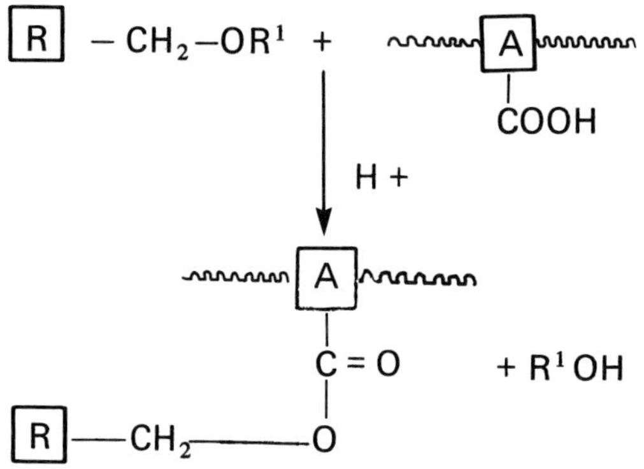

$$\boxed{R} - CH_2 - OR^1 \ + \quad \sim\sim\sim \boxed{A} \sim\sim\sim$$

$$\downarrow COOH$$

$$\downarrow H+$$

$$\sim\sim\sim \boxed{A} \sim\sim\sim$$

$$C = O \quad + R^1OH$$

$$\boxed{R} - CH_2 \text{------} O$$

Where R^1 is a short chain alcohol.

d) Chroman ring formation.
   A further reaction, Chroman ring formation, can occur where a drying oil is present in the alkyd and a phenolic resin is used as co-curing agent.

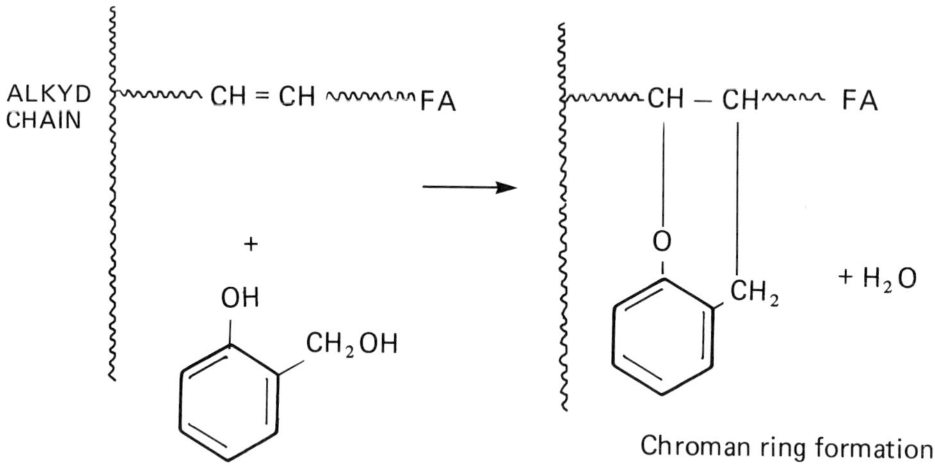

Chroman ring formation

A similar reaction has been reported to take place with melamine and urea-formaldehyde resins, provided they are in the enol form.

$$HOC_2N - C - N - CH_2OH \rightleftharpoons HOCH_2 N - C = N - CH_2OH$$

H   O   H   H   OH

Urea (keto form)   **Urea (enol form)**

The methylol groups on the amino resins also undergo self-condensation at the temperatures normally employed to cure these systems.

$$\{- O - CH_2-R - CH_2OH + HOCH_2 - R - CH_2 - O -\}$$

ALKYD CHAIN $-H_2O$ ALKYD CHAIN

$$\{- O - CH_2-R - CH_2 - O - CH_2 - R - CH_2 - O -\}$$

$- HCHO$

$$\{- O - CH_2-R - CH_2 - R - CH_2 - O -\}$$

(see Chapter on Amino and Phenolic Resins)

The presence of unsaturated oils, or their fatty acids, in the alkyd, also means that additional cross links can be formed by oxidation, as well as by the reactions outlined above.

This results in harder films being formed, for the same alkyd to amino resin blend ratio. The more expensive amino resin component can therefore be reduced, without reducing film hardness properties, if semi-drying oils are incorporated in the alkyd.

In practice however, the darker colours produced on overbake, by systems employing unsaturated oils or fatty acids, restricts their use.

It can be seen from the above that an important factor in alkyds intended for co-cure systems, is the number of residual hydroxyl and carboxylic acid groups present.

These alkyds are generally lower in molecular weight (before cure), than those used in air-drying systems. They are formulated and manufactured to rigidly controlled hydroxyl value and acid value specifications.

There is not the same amount of tolerance for variation in final resin constants as there is with other types. Departure from the required specification will almost certainly result in a sub-standard product on curing.

Typical acids values are in the range 15−25 mgKOH/g, while hydroxyl values can vary from 80−140 mgKOH/g depending on the intended end use.

The alkyd amino resin blend may be cured in the presence of a strong acid catalyst (e.g. paratoluenesulphonic acid) at temperatures ranging from 140−180°C.

When lower temperature cure is required, stronger acids, such as hydrochloric acid, or occasionally phosphoric acid, are used as catalysts.

Suitably formulated systems may be cured without the use of acidic catalysts.

Another method occasionally used for improving the cure time, or lowering the cure temperature, is to incorporate a strong acid onto the alkyd resin chain, after completion of the polycondensation stage. This can be undertaken by reacting tetrachlorophthalic, or tri-mellitic anhydride, onto some of the residual hydroxyl groups on the alkyd chain. Care must be exercised to ensure that only part esters are formed, otherwise cross-linking and gelation will result.

The process is usually carried out at 120−150°C, after polycondensation of the alkyd and before diluting with solvent.

It is important that only a small percentage of the available hydroxyl groups are treated in this way, since the cure reaction would be adversely effected if there were too few hydroxyl groups.

Alkyd resins react slowly with amino resins, even at room temperature, due to the presence of carboxylic acid groups on the alkyd. Hence resins must be formulated with shelf life, as well as cure rate in mind. Sometimes tertiary amines (e.g. tri-ethylamine) are added to the alkyd amino resin blends, to improve shelf-life. The amine temporarily removes the carboxylic acid group by forming an amine salt. On heating, the amine is lost from the system and the acid group is free to catalyse the reaction between the methylol and hydroxyl groups.

THE EFFECT OF OIL, OR FATTY ACID, ON THE PROCESSING CHARACTERISTICS OF AN ALKYD RESIN

There is a considerable amount of literature available from raw material suppliers detailing the effect of different raw materials on the film properties of the alkyd. This section will therefore be largely confined to the effect of raw materials on processing characteristics.

Drying Oils

Linseed and tung oils are easily heat polymerised and in the case of tung oil, gelation occurs readily at elevated temperatures. When carrying out alcoholysis on linseed oil it is important to limit the temperature to 245−250°C and to restrict the time taken to obtain alcoholysis by using an effective catalyst. Failure to control either parameter can result in considerable inter-batch variation caused by the presence of varying amounts of heat polymerised oil.

Tung oil is seldom used on its own. A blend of tung oil with other drying oils (i.e. linseed or dehydrated castor oil) is almost always used. When it is present in a formulation, it is normal to restrict the reaction temperature to a maximum of 240°C (a temperature of 220°C being preferred).

Dehydrated castor oil and its fatty acid are frequently used in alkyd resin and modified alkyd resin manufacture. Often the raw castor oil is dehydrated 'in situ' in the presence of the polyol and phthalic anhydride with the phthalic anhydride also acting as dehydration catalyst.

As previously discussed (see glyceride oils) dehydration requires a minimum temperature of 265−270°C and an inert gas sweep to remove the water formed during the reaction.

Under these conditions, losses of volatile polyols and phthalic anhydride can be large and this must be taken into consideration when formulating this type of alkyd.

This is particularly relevant when employing mixed polyols, where one species is considerably more volatile than the other, e.g. glycerol pentaerythritol mixtures.

The course of the dehydration is similar to that discussed under glyceride oils. It should be remembered, however, that polycondensation is also occuring simultaneously and once dehydration has been completed, the viscosity which has fallen steadily to this point will now increase rapidly. Once dehydration is complete, it is normal practice to cool the reactants to a lower temperature, to allow better control of the subsequent polycondensation reaction.

Semi-Drying Oils and Fatty Acids

Soya bean oil and sunflower oil are interchangeable in most formulations. Both oils tend to bleach at alcoholysis temperatures and colours are generally below 7 Gardner. Tall oil fatty acid can be used as a replacement for soya bean and sunflower fatty acids, but the polycondensation reaction exhibits a different acid value viscosity relationship, when tall oil fatty acid and glycerol are used to replace soya bean oil.

Non-Drying Oils and Fatty Acids

Alkyds formulated on saturated oils and fatty acids are designed to give very low colours. Colour can be preserved by the inclusion of small quantities of tri-phenyl phosphite.

However, the presence of tri-phenyl phosphite, interfers with the alcoholysis, and so it must be added after this stage of the process has been completed.

Tri-glyceride oils, or their derived fatty acids are not used exclusively in non-drying coatings. Other acid containing species, not strictly classed as oils or oil derived fatty acids, may also be used. These are discussed under modifiers. They include such materials as heptanoic acid and rosin.

Polyols

The most frequently used polyols are:

Glycerol

$$CH_2 - OH$$
$$|$$
$$CH \ - OH$$
$$|$$
$$CH_2 - OH$$

Pentaerythritol

$$CH_2OH$$
$$|$$
$$HOCH_2 \longrightarrow C \longrightarrow CH_2OH$$
$$|$$
$$CH_2OH$$

Trimethylol Propane

$$CH_2OH$$
$$|$$
$$CH_3 - CH_2 - C \longrightarrow CH_2OH$$
$$|$$
$$CH_2OH$$

The replacement of glycerol by pentaerythritol, results in films with improved hardness. However, the increased functionality of pentaerythritol, compared to glycerol, (4 instead of 3) leads to a more cross-linked polymer, unless the overall functionality of the system is modified.

Glycerol may be replaced by equimolar quantities of ethylene glycol and pentaerythritol, to give a blend with a similar overall functionality to that of glycerol.

However, although still practised, this technique is declining in popularity, since it is difficult to prevent excessive losses of ethylene glycol during manufacture without considerably extending the process time.

It is now becoming more usual to employ pentaerythritol glycerol blends, in short oil length alkyds and to use monofunctional acid species as 'chain stoppers' or alternatively use a large hydroxyl excess.

The use of pentaerythritol or tri-methylol propane, in place of glycerol, allows the fusion process to be operated, without excessive polyol losses being encountered.

When using tri-methylol propane, it should be remembered that all hydroxyl groups are of equal reactivity, (unlike glycerol which has 1 secondary and 2 primary hydroxyl groups). Therefore the viscosity acid value relationship of similar formulations with glycerol and tri-methylol propane, will be different.

Di-basic Acids

The main di-basic acid, used in alkyd resin manufacture is phthalic anhydride.

Isophthalic acid is sometimes used, where improved film hardness and chemical resistance are required.

and sometimes terephthalic acid is also used for the same reason.

$$
\begin{array}{c}
O \\
\parallel \\
C-OH
\end{array}
$$

Terephthalic acid is not soluble in the raw materials normally employed in alkyd resin manufacture and can only be reacted into the system by prolonged heating at temperatures in excess of 240°C. For this reason it is often added in the form of the more soluble dimethyl terephthalate derivative, which undergoes ester interchange (transesterification) with the polyols liberating methanol, which has to be removed under distillation.

Isophthalic acid is more soluble than terephthalic acid but, nonetheless presents processing problems and prolonged heating at elevated temperatures is often required.

Isophthalic acid is often used in blends with phthalic anhydride. The isophthalic is added to the reactor following alcoholysis, when there is a large excess of hydroxyl groups available for reaction. The mixture is held at 240°C or higher under fusion process, until the isophthalic has reacted into the system, as evidenced by the acid value and the clarity of a cold film of a sample from the reactor.

Once the isophthalic has reacted into the system, the phthalic anhydride part of the acid charge is added and the polycondensation reaction then continued, either by fusion or solvent process techniques.

Phthalic anhydride does not suffer from major solubility problems since the anhydride ring readily opens and reacts with the polyol to form a half-ester at temperatures as low as 150°C. Once the half-ester has been formed the phthalic anhydride has effectively been solubilised.

Modifiers

This section deals with materials that are added in small quantity, usually less than 5%. Major modifications are discussed later in the section on modified alkyds.

MONO-BASIC ACIDS (Excluding oil derived fatty acids)

A wide range of mono-basic acids are employed commercially. Only some of these are mentioned in this section.

Where the alkyd is of the oxidising type, the mono-basic acid content has to be almost exclusively based on drying or semi-drying fatty acids. Although in high functionality systems (e.g. when pentaerythritol is used) small amounts of benzoic acid are employed and act as an additional chain terminator.

Long oil length oxidising alkyds are frequently modified with rosin (or rosin esters) for use in letterpress and lithographic ink systems where rapid surface dry is required.

The inclusion of rosin (essentially abietic acid):

acts to improve the drying speed and 'set' of the alkyd.

Where rosin is employed, it is necessary to carry out the alcoholysis stage prior to addition of the rosin to avoid interference, not only with mono-glyceride formation, but also to the testing methods used to determine the degree of alcoholysis during manufacture. (Rosin modification is discussed in detail in the section on oleoresinous varnishes).

Non-oxidising alkyds are frequently made, using short chain acids, such as heptanoic acid, as a partial or even complete replacement for glyceride oils or fatty acids.

Synthetic coconut oil fatty acid substitute, produced by blending various saturated mono-basic acid species, is commonly used in amino cured alkyd systems.

DI-BASIC ACIDS

The major materials used as di-basic acid and modifiers are dimer acid (dimerised linoleic acid) and maleic anhydride or fumaric acid. Maleic anhydride and fumaric acid can be discussed as one, under maleic anhydride, since their effects are similar.

Dimer acid is frequently used in small amounts ca 2−4% on the solid resin as a functionality modifier to chain extend alkyds, effectively increasing the product viscosity. The effect of using small amounts of dimer acid is that an increase in the overall solubility of the resin results and thus a modified dilution curve can be obtained.

Maleic anhydride is frequently used in small amounts, usually less than 1% of the solid resin to increase resin functionality and hence product viscosity at a given degree of esterification. The maleic anhydride acts by adding across the double bonds of the unsaturated fatty acids present (see chapter on glyceride oils) effectively forming a trifunctional acid which acts as a site for chain branching. Great care must be taken when using this technique, otherwise gelation can occur.

TYPICAL ALKYD RESIN FORMULATIONS

The following formulations are intended as examples of typical commercially available alkyd resin types. Brief process details are included which will enable the reader to prepare these resins for use in the accompanying application formulations.

Oxidising Alkyds

Long Oil Length Linseed-Penta Alkyd

FORMULATION

Linseed oil (first grade)	49.85
Pentaerythritol	8.61
Lithium hydroxide (alcoholysis catalyst)	0.01
Phthalic anhydride	17.07
White spirit (diluent)	24.46
	100.00

PROCESS

A two stage process is used with alcoholysis followed by polycondensation, using the fusion process technique.

1. **Alcoholysis:** the linseed oil, pentaerythritol and lithium hydroxide are heated together under an inert gas atmosphere at 245–250°C. Alcoholysis is deemed complete when a sample of the reaction mixture has a minimum tolerance at 25°C of 1 part sample ex-reactor to 3 parts 74. op methanol.

2. **Polycondensation:** on completion of alcoholysis the reaction mixture is cooled to 180°C and phthalic anhydride added. The reactants are reheated to 240°C and held at this temperature under an inert gas sweep, to remove water from the reaction. The course of the esterification is monitored by measurement of acid value and viscosity. The reaction is quenched by cooling to below 180°C and diluting with white spirit, when an acid value below 10 mgKOH/g and a viscosity (measured at a temperature of 25°C on a sample diluted to 75% in white spirit) of 40–50 poise is obtained.

FINAL RESIN CONSTANTS

Colour	9 Gardner (maximum)
Viscosity at 25°C	40–60 poise (Gardner tubes Z_7–Z_1)
Acid value	10 mgKOH/g (maximum)
Non-volatile content	74–76%
Oil length	68%

This alkyd can be used for general purpose air drying points for exterior use, where good durability is required.

It can also be used in formulations for lithographic inks.

Long Oil Length DCO-Glycerol Alkyd

FORMULATION

Dehydrated castor oil fatty acid	37.84
Phthalic anhydride	16.45
Glycerol	12.25
Xylene (reflux solvent)	3.98
White spirit (diluent)	19.75
Xylene (diluent)	9.73
	100.00

PROCESS

Polycondensation is carried out at 230°C using xylene to aid removal of water of reaction. Since fatty acid rather than oil is used, no alcoholysis stage is required. Control is for an acid value below 6 mgKOH/g and a viscosity (determined on a sample diluted to 65% in 1 part xylene to 2 parts white spirit) of 7–12 poise at 25°C. Once the desired viscosity and acid value are achieved the reaction is quenched by cooling below 180°C and diluting with solvent. At 230°C approximately 6% solvent is present in the reactor and this should be taken into consideration when preparing a 65% solution for viscosity determination.

FINAL RESIN CONSTANTS

Colour	5 Gardner (maximum)
Viscosity at 25°C	7–12 poise (Gardner tubes U + to X–)
Acid value	6 mgKOH/g (maximum)
Non-volatile content	64–66%

APPLICATIONS

This type of resin is used for non-yellowing air drying varnishes for paint and ink formulations.

Long Oil Length Soya-Pentaerythritol Alkyd

FORMULATION

Alkali refined soya bean oil	43.85
Pentaerythritol	9.57
Phthalic anhydride	16.94
Fumaric acid	0.29
Xylene (reflux solvent)	2.84
White spirit (diluent)	26.51
	100.00
Caustic Soda (5% aq. solution)	0.05 (alcoholysis catalyst)

PROCESS

A two stage process is used and this involves alcoholysis followed by polycondensation using xylene to azeotropically remove water formed during the reaction.

1. **Alcoholysis:** soya bean oil, pentaerythritol and caustic soda solution are heated together at 245°C under an inert gas blanket. Alcoholysis is deemed complete when a sample of the reactants has a minimum tolerance of 1 part sample to 3 parts 74 o.p. methanol at 25°C.

2. **Polycondensation:** on completion of alcoholysis, the reactants are cooled to 180°C and the pentaerythritol, phthalic anhydride, fumaric acid and xylene (reflux solvent) are added. The presence of fumaric acid increases the molecular complexity of the resin, and leads to a product with a higher viscosity. Esterification is carried out at 260°C, using the solvent process technique for an acid value below 10 mgKOH/g and a viscosity at 25°C of 25–30 poise (determined on a 70% solution in white spirit). The reaction is quenched by cooling below 180°C and diluting with white spirit.

FINAL RESIN CONSTANTS

Colour	8 Gardner (maximum)
Viscosity (at 25°C)	25–30 poises (Gardner tubes Z_1 – Z_2)
Acid value	10 mgKOH/g (maximum)
Non-volatile content	69 – 71%

APPLICATIONS

This alkyd is used as the basis for air drying decorative paints. It gives fast drying films with good gloss and colour retention. It can be used for pastel shades and whites.

Short Oil Length Dehydrated Castor Oil-Pentaerythritol Alkyd employing P-Tert-Butyl Benzoic Acid Modification

FORMULATION

Dehydrated castor oil	23.88
Pentaerythritol (1)	7.15
Lithium hydroxide (alcoholysis catalyst)	0.01
Pentaerythritol (2)	3.07
Phthalic anhydride	15.00
p-tert-butyl benzoic acid	12.12
Xylene (reflux solvent)	5.06
Xylene (diluent)	33.71
	100.00

PROCESS

1. **Alcoholysis:** the dehydrated castor oil, together with pentaerythritol (1) and lithium hydroxide are heated to 240−250°C under an inert gas blanket. Only part of the total pentaerythritol content is present during alcoholysis. This is to prevent excess unreacted pentaerythritol from interfering with the tolerance test. Alcoholysis is deemed complete when a minimum tolerance at 25°C of 1 part of the reaction mixture to 3 parts of methanol is obtained.

2. **Polycondensation:** on completion of alcoholysis the reactants are cooled to 200°C and the remaining pentaerythritol (2) is added. Cooling is continued to below 180°C.

Then the phthalic anhydride and p-tertiary butyl benzoic acid are added and the reactor heated to 220°C.

The p-tertiary-butyl-benzoic acid is present to reduce the functionality of the system. The mono-functional acid reacts with the hydroxyl groups of the pentaerythritol, preventing further chain extension. In this way, the molecular weight (and hence viscosity) of the resin are effectively controlled. P-tertiary-butyl-benzoic acid is frequently used in preference to other mono-functional acids because of the improved film hardness properties which it confers, compared to the properties obtained from other mono-functional short chain acids. At this temperature the xylene reflux solvent is added and esterification carried out, using the solvent process technique. The course of the reaction is monitored by acid value and viscosity determination. It is controlled for an acid value in the range of 15−20 mgKOH/g (on calculated solid resin) and a viscosity at 25°C of 95−105 poises (determined on a 60% solution of the resin in xylene).

Once the desired degree of reaction has been achieved, it is quenched by cooling below 180°C and diluting with xylene.

FINAL RESIN CONSTANTS

Colour	6 Gardner maximum
Viscosity (at 25°C)	100 – 140 poises (Gardner tubes $Z_5 - Z_6$)
Acid value	15 – 20 mgKOH/g
Non-volatile content	59 – 61%
Oil length	41%

APPLICATIONS

This alkyd can be utilised in industrial stoving finishes and rapid air-drying finishes. The rapid air-drying is coupled with excellent non-yellowing properties (conferred by the D.C.O.) and this type of resin can be used in white and pastel shade finishes.

Short Oil Length D.C.O.F.A. — Glycerol Alkyd

FORMULATION

Dehydrated castor oil fatty acid	17.68
Glycerol	13.10
Phthalic anhydride	20.46
Xylene (reflux solvent)	4.53
Xylene (diluent)	44.23
	———
	100.00

PROCESS

This resin is prepared by the fatty acid process. The D.C.O. fatty acid, glycerol and phthalic anhydride, together with xylene (reflux solvent) are charged to the reactor and heated to 200°C. Polycondensation is carried out using the solvent process technique. The course of the reaction is monitored by acid value and viscosity determination and the reaction is quenched when an acid value of 20–25 mgKOH/g and viscosity (measured at 25°C on a 50% solution of the resin in xylene) of 40–50 poises is obtained.

FINAL RESIN CONSTANTS

Colour	6 Gardner (maximum)
Viscosity at 25°C	40 – 50 poises (Gardner tubes $Z_2 - Z_3 +$)
Acid value	20 – 25 mgKOH/g
Non-volatile content	49 – 51%
Oil length	42%
Hydroxyl value	115 – 125 mgKOH/g

APPLICATIONS

This type of alkyd can be used on its own, as a medium for stoving enamels and primers. Alternatively it can be cured in conjunction with an amino (urea or melamine) resin.

This resin gives films with excellent gloss and colour retention. It is compatible with a wide range of amino resin systems. Stoving schedules of 120°C for 30 minutes are typically used for alkyds of this type.

Medium Oil Length Tall Oil Fatty Acid — Glycerol Alkyd

FORMULATION

Tall oil fatty acid	22.33
Glycerol	11.53
Phthalic anhydride	17.96
Xylene (reflux solvent)	2.54
White spirit (diluent)	45.64
	100.00

PROCESS

Since fatty acids are used, no alcoholysis stage is required. Polycondensation is brought about at 240°C, using xylene, azeotropically to remove ester formed during the reaction. The reaction is quenched, once an acid value of 15 – 20 mgKOH/g and a viscosity at 25°C (determined on a 50% solution of the sample, i.e. xylene) of 3.0 – 4.0 poises is obtained.

It should be noted that at 240°C the sample ex-reactor contains approximately 4.5% solvent. This must be taken into account when the sample is diluted to 50% for viscosity measurements.

FINAL RESIN CONSTANTS

Colour	6 Gardner (maximum)
Viscosity at 25°C	20.0 – 30.0 poises (Gardner tubes Y + Z_1 +)
Acid value	15 – 20 mgKOH/g
Non-volatile content	49 – 51%

APPLICATIONS

This resin can be used in clear or pigmented forced drying or air drying systems. Colour retention after stoving is good.

Short Oil Length Dehydrated Castor Oil — Glycerol Alkyd Modified with Gum Rosin

FORMULATION

Raw castor oil (first pressings)	16.75
Glycerol	10.84
Phthalic anhydride	18.23
Gum rosin	5.91
Xylene (diluent)	48.27
	100.00

PROCESS

The resin is made using a two stage process. The castor oil is dehydrated 'in-situ' at 270°C, followed by polycondensation at 200°C.

1. **Dehydration:** the reactants, with the exception of the xylene are heated to 270°C under an inert gas blanket. The course of the dehydration is followed by monitoring the viscosity. As dehydration proceeds the viscosity decreases, reaching a minimum value when the dehydration is almost complete. Once this minimum viscosity has been reached and an increase is noted, the reactants are immediately cooled to 200°C.

2. **Polycondensation:** is carried out at 200°C using the fusion process technique and the reaction is quenched once a viscosity at 25°C of 4.5 – 5.5 poises (determined on a sample diluted to 50% with xylene) and acid value of 15 – 20 mgKOH/g is achieved.

FINAL RESIN CONSTANTS

Colour	12 – 13 Gardner
Viscosity at 25°C	4.5 – 5.5 poises (Gardner tubes Q to T)
Acid value	15 – 20 mgKOH/g
Non-volatile content	48 – 52%

APPLICATIONS

This resin is particularly useful in areas where rapid cure is required. It can be used alone in air-drying finishes or in general purpose stoving finishes in combination with 15% by weight of a melamine or 20% by weight of a urea resin.

Short Oil Length Soya-Tri-Methylol Propane Stoving Alkyd

FORMULATION

Refined soya bean oil	18.200
Tri-methylol propane	14.170
Litharge (alcoholysis catalyst)	0.007
Phthalic anhydride	18.770
Xylene (reflux solvent)	3.120
Xylene (diluent)	45.740
	100.000

PROCESS

Manufacture is carried out in two stages.

1. **Alcoholysis:** soya bean oil, tri-methylol propane and litharge are heated together at 245°C under an inert gas blanket. Alcoholysis is deemed to be complete when a tolerance of 1 part sample (ex-reactor) to 3 parts 74 o.p. methanol is obtained at 25°C.

2. **Polycondensation:** the alcoholysis product is cooled to 180°C and phthalic anhydride and xylene (reflux solvent) are added. The reactants are then heated at 225°C and esterification is brought about using the solvent process technique to aid in removing water of reaction. The reaction is quenched by cooling below 180°C and adding the diluent once a viscosity (50% in xylene) of 7.5 – 9.5 poises at 25°C and an acid value below 8 mg/KOH/g have been obtained. Approximately 6% xylene will be present in the reactor at a solvent process temperature of 225°C and this must be taken into account when calculating the dilution prior to viscosity determination.

FINAL RESIN CONSTANTS

Colour	6 Gardner (maximum)
Viscosity	8.0 – 9.5 poises (Gardner tube V to V +)
Acid value	8 mgKOH/g (maximum)
Non-volatile content	49 – 51%
Hydroxyl value	70 – 80 mgKOH/g
Oil length	37%

APPLICATIONS

This resin is used in stoving enamels either on its own or with amino resins. The alkyd has excellent flexibility and good alkali resistance. Typical stoving schedule is 35 minutes at 140°C.

Non-Oxidising Alkyds

Long Oil Length Castor Oil Plasticising Alkyd

FORMULATION

Castor oil fatty acid	53.59
Sebacic acid	15.00
Dimer acid (dimerised linoleic acid)	17.58
Glycerol	13.83
	100.00

PROCESS

The castor oil fatty acid is heated to 150°C and the sebacic acid fed into the fatty acid slowly over a 30 minute period. Once the sebacic acid has dissolved, the dimer acid and glycerol are added and the temperature is raised to 240°C.

Polycondensation is carried out using the fusion process technique and an inert gas sweep is maintained to aid the removal of water formed during the reaction. Esterification is continued until an acid value below 10 mgKOH/g and a viscosity of 100 – 125 poises are obtained.

FINAL RESIN CONSTANTS

Colour	5 Gardner (maximum)
Viscosity of 25°C	100 – 150 poises (Gardner tubes $Z_5 - Z_6$)
Acid Value	10 mgKOH/g (maximum)

APPLICATIONS

This resin has excellent colour retention and is completely compatible with nitrocellulose. It is used as a plasticiser in nitrocellulose finishes.

Short Oil Length Castor Oil Plasticising Alkyd

FORMULATION

Hydrogenated castor oil	21.63
Glycerol	5.95
Phthalic anhydride	25.42
Xylene (diluent)	47.00
	100.00

PROCESS

No alcoholysis is necessary because of the presence of hydroxyl groups on the castor oil. The reactants are heated together to 245°C. Polycondensation is carried out using the fusion process technique.

The reaction is quenched by cooling below 180°C and diluting with xylene when a viscosity at 25°C of 15 – 20 poises (on a sample diluted to 50% in xylene is obtained).

FINAL RESIN CONSTANTS

Colour	3 Gardner (maximum)
Viscosity	15 – 20 poises (Gardner tubes Y to Z)
Acid value	28 – 35 mgKOH/g
Non-volatile content	49 – 51%
Oil length	46%

APPLICATIONS

This resin is designed for use as a plasticising alkyd in colour retentive stoving finishes and nitrocellulose lacquers and exhibits excellent colour retention at high temperatures.

It confers good flexibility and gas checking resistance and has good surface hardness properties. In combination with urea or melamine resins it acts as a plasticising resin to improve adhesion and flexibility while maintaining high temperature colour resistance and it improves the acid and alkali resistance of the system.

Short Oil Length Coconut Oil — Glycerol, Pentaerythritol Alkyd

FORMULATION

White distilled coconut oil fatty acid	17.04
Phthalic anhydride	20.65
Glycerol	6.71
Pentaerythritol	7.40
Xylene (reflux solvent)	8.35
Xylene (diluent)	44.85
	100.00

PROCESS

The reactants are heated together with xylene (reflux solvent) to 220°C. Polycondensation is carried out using the solvent process technique and the reaction is quenched when an acid value of 25–30 mgKOH/g and a viscosity at 25°C (determined on a 50% solution of the resin in xylene) of 35–40 poises are achieved. The solvent content of the reactants at 220°C is approximately 6.5% and this should be taken into consideration when preparing the sample for viscosity determination.

FINAL RESIN CONSTANTS

Colour	3 Gardner (maximum)
Viscosity	35–40 poises (Gardner tubes $Z_2 - Z_2 +$)
Acid value	25–30 mgKOH/g
Non-volatile content	49–51%
Hydroxyl value	100–110 mgKOH/g

APPLICATIONS

This type of resin is co-cured with amino resins. It can be used for non-yellowing stoving coatings particularly for whites and pastel shade enamels for machinery, automobile and appliance finishes. The film properties are to a large extent governed by the ratio of alkyd to amino resin employed and the stoving schedule used.

Short Oil Length Coconut Oil — Glycerol, Pentaerythritol Alkyd

FORMULATION

White distilled coconut oil fatty acid	17.11
Phthalic anhydride	20.75
Glycerol	8.99
Pentaerythritol	4.93
Xylene (reflux solvent)	3.35
Xylene (diluent)	44.84
	100.00

PROCESS

The reactants and xylene (reflux solvent) are heated to 220°C and polycondensation is carried out, using the solvent process technique. The reaction is quenched when an acid value of 20−25 mgKOH/g and a viscosity at 25°C, (measured on a sample diluted to 50% solid content in xylene) of 36−46 poises are obtained. The solvent content of the resin at a process temperature of 220°C is approximately 3.5% and this should be taken into consideration when diluting a sample to 50% solids content prior to determining the viscosity.

FINAL RESIN CONSTANTS

Colour	3 Gardner (maximum)
Viscosity	36 − 46 poises (Gardner tubes $Z_2 - Z_3$)
Acid value	20 − 25 mgKOH/g
Non-volatile content	49 − 51%
Hydroxyl value	105 − 115 mgKOH/g
Oil length	40%

APPLICATIONS

This resin is cured (as with the previous alkyd) in conjunction with amino resins. The difference between this and the preceding formulation is the increased ratio of glycerol to pentaerythritol which makes films from this resin slightly softer and more flexible than those obtained with the previous formulation.

Short Oil Length Coconut Oil — Glycerol Alkyd Modified with Dimer Acid

FORMULATION

White distilled coconut oil fatty acid	15.16
Dimer acid (dimerised lineoleic acid)	7.92
Phthalic anhydride	16.66
Glycerol	12.01
Xylene (reflux solvent)	3.35
Xylene (diluent)	44.90
	100.00

PROCESS

The reactants are heated together with xylene (reflux solvent) to 220°C. Polycondensation is brought about using the solvent process technique to aid water removal. The reaction is quenched when an acid value of 20−25 mgKOH/g and a viscosity (at 25°C) of 1.8−2.2 poises (determined on a sample diluted to 50% in xylene) are obtained. Solvent content at 220°C is approximately 6.5% and this should be taken into consideration when preparing a sample for viscosity determination.

FINAL RESIN CONSTANTS

Colour	6 Gardner (maximum)
Viscosity at 25°C	1.8−2.2 poises (Gardner tubes H + to I−)
Acid value	20−25 mgKOH/g
Non-volatile content	49−51%
Hydroxyl value	90−100 mgKOH/g
Oil length	40%

APPLICATIONS

This resin is used in conjunction with amino resins in stoving finishes. It is particularly suitable for coating substrates of copper or copper alloys where good adhesion is required and discolouration of the clear lacquer must be kept to a minimum. The reduced tendency of this resin to attack copper is of particular value in electrical insulating varnishes.

Short Oil Length Coconut Oil Stoving or Plasticising Alkyd

FORMULATION

Coconut oil fatty acid	5.51
Stearic acid	8.82
Glycerol	14.53
Phthalic anhydride	23.01
Xylene (reflux solvent)	3.33
Xylene (diluent)	44.80
	100.00

PROCESS

The coconut fatty acid is heated to 100°C and the stearic acid added. Once the stearic acid has dissolved, the glycerol, phthalic anhydride and xylene (reflux solvent) are added and the reactor heated to 220°C. Polycondensation is carried out at 220°C using the solvent process technique to aid removal of water of reaction. The esterification is terminated by cooling the reactants below 180°C and adding xylene (diluent) when a viscosity at 25°C, (determined on a 50% solution of the resin in xylene) of 25−35 poises and acid value 18−25 mgKOH/g are obtained. The solvent content of the reactor at 220°C is approximately 6.5% and this must be taken into consideration when preparing the sample for viscosity determination.

FINAL RESIN CONSTANTS

Colour	5 Gardner (maximum)
Viscosity at 25°C	25 − 35 poises (Gardner tubes $Z_1 − Z_2 −$)
Acid value	18 − 25 mgKOH/g
Non-volatile content	49 − 51%
Hydroxyl value	120 − 130 mgKOH/g
Oil length	31%

APPLICATIONS

This resin can be used as a plasticising alkyd with nitrocellulose lacquers or as a stoving enamel in conjunction with amino resins. The resin has excellent adhesion and toughness together with excellent colour retention and resistance to blooming. The resin has excellent pigment wetting properties which allows a high gloss stoving finish to be achieved in conjunction with amino resins.

ALKYD RESINS FORMULARY

The following formulations have been taken from the literature. Where possible they have been selected to incorporate the resins detailed in the preceding pages.

Oxidising Types

WHITE GLOSS ENAMEL FOR BRUSH APPLICATION

	Rutile TiO$_2$	23.40
	Long oil linseed-penta alkyd (75% non-volatile content)	51.90
Driers	Cobalt naphthenate (6% Co)	0.32
	Lead naphthenate (24% Pb)	0.79
	Methyl Ethyl ketoxime	0.08
	White spirit	23.51
		100.00

RAPID AIR DRYING GLOSS INK — RED

	Toluidene red	7.76
	60% oil length linseed/ tung oil — penta alkyd (55% non-volatile content)	70.50
	Cobalt naphthenate (6% Co)	0.48
Driers	Lead naphthenate (24% Pb)	0.81
	Calcium naphthenate (5% Ca)	1.54
	Methyl ethyl ketoxime	0.19
	Solvent SBP 6	18.72
		100.00

RED METAL DECORATING INK

	70% Oil length D.C.O. alkyd (non-volatile content)	44.2
	Cyclic ketone resin (e.g. MS2 ex-Laporte Chemicals)	14.7
	High boiling hydrocarbon solvent (e.g. fusil oil)	14.7
	Rutile TiO_2	3.9
	Red pigment (or red pigments blends)	19.4
Driers	Cobalt naphthenate (6% Co)	0.4
	Zinc naphthenate (8% Zn)	2.7
		100.00

Stoving schedule 15 minutes at 150°C

MAGENTA LITHOGRAPHIC INK

Red pigment (or red pigments blend)	31.54
70% Oil length D.C.O. alkyd (100% non-volatile content)	47.32
Surface treated calcium carbonate	12.62
Alkali refined linseed oil	7.89
Cobalt naphthenate (6% Co)	0.39
Methyl ethyl ketoxime	0.24
	100.00

BLACK LETTERPRESS INK

70% Oil length D.C.O. alkyd (100% non-volatile content)	63.39
Carbon black	13.58
Prussian blue	9.62
Alkali refined linseed oil	12.88
Cobalt naphthenate (6% Co)	0.53
	100.00

WHITE BRUSHING ENAMEL — NON-YELLOWING

	Rutile TiO$_2$	26.60
	Long oil soya-penta alkyd (70% non-volatile content)	50.74
Driers	Cobalt naphthenate (6% Co)	0.29
	Lead naphthenate (24% Pb)	0.24
	Calcium naphthenate (4% Ca)	0.89
	White spirit	20.74
		100.00

RED BRUSHING MACHINERY FINISH

	Permanent red R	8.60
	Soya leeithin	0.09
	Long oil soya-penta alkyd (60% non-volatile content)	71.50
Driers	Cobalt naphthenate (6% Co)	0.36
	Lead naphthenate (24% Pb)	0.88
	Calcium Naphthenate (4% Ca)	0.86
	Methyl ethyl ketoxime	0.10
	White spirit	17.61
		100.00

GENERAL PURPOSE WHITE STOVING ENAMEL

Rutile TiO$_2$	20.50
Approx. 40% oil length short oil D.C.O. glycerol alkyd (non-volatile content)	42.80
Butylated urea resin (e.g. Beetle 610 ex-BIP Chemicals)	21.40
n-butyl acetate	1.90
n-butanol	3.60
Xylene	9.80
	100.00

Stoving schedule 30 minutes at 160°C
45 minutes at 140°C

HIGH GLOSS STOVING ENAMEL

Rutile TiO_2	25.78
Approx. 40% oil length D.C.O. glycerol alkyd (non-volatile content 50%)	43.31
Urea resin (e.g. Beetle 610 ex-BIP Chemicals)	12.18
Xylene	14.05
n-butanol	4.68
	100.00

Stoving schedule 30 minutes at 120°C

QUICK AIR DRYING GREY GLOSS FINISH

Rutile TiO_2	17.20
Ferrite yellow	1.50
Lamp black	0.56
Short oil D.C.O. glycerol alkyd (non-volatile content)	59.50
Cobalt naphthenate (6% Co)	0.32
Methyl ethyl ketoxime	0.33
Xylene	20.50
	100.00

WHITE LOW TEMPERATURE STOVING FINISH

Rutile TiO_2	22.60
Medium oil length tall oil glycerol alkyd (non-volatile content 50%)	54.30
Melamine resin	5.50
Driers { Cobalt naphthenate (6% Co)	0.36
Lead naphthenate (24% Pb)	1.13
Calcium naphthenate (5% Ca)	1.08
Methyl ethyl ketoxime	0.11
White spirit	14.92
	100.00

Stoving schedule
30 minutes at 80°C

BLUE STOVING ENAMEL

Rutile TiO_2	14.50
Phthalocyanine blue	1.45
Short oil length D.C.O. glycerol alkyd (non-volatile content 50%)	61.70
Butylated urea resin (e.g. Beetle 610 or 640 ex-BIP Chemicals)	8.80
Xylene	13.55
	100.00

Stoving schedule 30 minutes at 135°C

AIR DRYING TRACTOR BLUE GLOSS FINISH

	Rutile TiO_2	5.04
	Prussian blue	1.57
	Phthalocyanine blue	0.64
	Lamp black	0.04
	Lemon chrome	0.08
	Soya lecithin	0.20
	Medium/short oil length tall oil FA alkyd (60% non-volatile content)	63.00
Driers	Cobalt naphthenate (6% Co)	0.38
	Lead naphthenate (24% Pb)	1.59
	Methyl ethyl ketoxime	0.01
	Xylene	27.45
		100.00

MEDIUM TONED QUICK-SETTING BLACK INK

	Carbon black	18.00
	Bronze blue	5.00
	Reflex blue/oil paste	5.00
	Long oil linseed/tung alkyd (70% non-volatile content)	58.00
	1.5 poises linseed stand oil	10.00
	Spindle oil	3.00
Driers	Lead naphthenate (24% Pb)	0.40
	Manganese naphthenate (6% Mn)	0.30
	Cobalt naphthenate (6% Co)	0.30
		100.00

Non-Oxidising Types

GENERAL PURPOSE WHITE STOVING ENAMEL

Rutile TiO_2	27.60
Hydrogenated castor oil alkyd (50% non-volatile content)	47.90
Butylated melamine resin (e.g. Beetle BE615 ex-BIP Chemicals)	17.27
Xylene	5.43
n-butanol	1.80
	100.00

Stoving schedule 30 minutes at 120°C
or 20 minutes at 150°C

NITROCELLULOSE POLISHING LACQUER

Short oil coconut stearic Acid alkyd (50% non-volatile content)	22.50
½ second nitrocellulose (70% damped in isopropanol)	13.20
Xylene	5.5
Di-octylphthalate	4.5
Thinners	42.90
Phenolic resin (50% in xylene)	11.40
	100.00

NITROCELLULOSE WOOD LACQUER

Short oil coconut stearic acid alkyd diluted in toluene (45% non-volatile content)	16.80
½ second nitrocellulose (70% damped in isopropanol)	15.99
Di-isobutyl phthalate	2.82
Toluene	22.29
n-butanol	7.43
Butyl acetate	24.76
Ethyl acetate	7.43
Ethylene glycol mono-ethyl ether (Cellosolve ex-Union Carbide)	2.48
	100.00

BLUE STOVING ENAMEL

Rutile TiO$_2$	13.00
Phthalocyanine blue	1.40
Short oil castor glycerol alkyd (45% non-volatile content)	50.70
Butylated melamine resin (Beetle 645 ex-BIP Chemicals)	20.20
Zinc naphthenate (8% Zn)	0.60
1% Silicone oil in xylene	13.00
	100.00

Stoving schedule 30 minutes at 150°C
20 minutes at 160°C

MODIFIED ALKYD RESINS

No single polymer type exhibits all the properties desired of a surface coating. Every polymer has at least one property that is less than ideal and commercial coatings are formulated to achieve a balance between the advantageous and disadvantageous properties of a base polymer system.

Combinations of different polymer types are frequently used to obtain a coating that combines the more desirable properties of individual polymer types. One of the great advantages of alkyd resins is the ease with which they can be chemically modified, by reaction with other polymers resulting in coatings combining the properties of both the alkyd and the modifying polymer.

There are several ways to bring these modifications about and this section deals with the more commercially important methods and polymer types.

VINYLATED ALKYD RESINS

A coating which combines the good application and wetting properties of the alkyd with the strength, weathering and chemical resistance of an acrylic or vinyl homopolymer, has obvious advantages. Blends of alkyd and vinyl or acrylic homopolymer are not readily compatible, partly due to the large difference in molecular weight of these types of polymer (e.g. alkyd molecular weights typically below 10,000 and the vinyl homopolymer molecular weights typically over 100,000).

Reduction in the molecular weight of the vinyl homopolymer to a level where compatibility with an alkyd can be achieved, also leads to a diminution in mechanical and weathering properties, since these are of course a function of molecular weight.

Where the vinyl homopolymer is of a thermosetting type, the reduction in molecular weight, necessitated by compatibility factors, has a less drastic effect on the final film properties. However, chemical modification is still preferred to straight blends.

Vinyl monomers can be introduced into the alkyd chain via the unsaturation of the fatty acid.

Where the fatty acid is conjugated, copolymerisation can proceed by a mechanism similar to that described in the chapter on addition polymerisation.

178

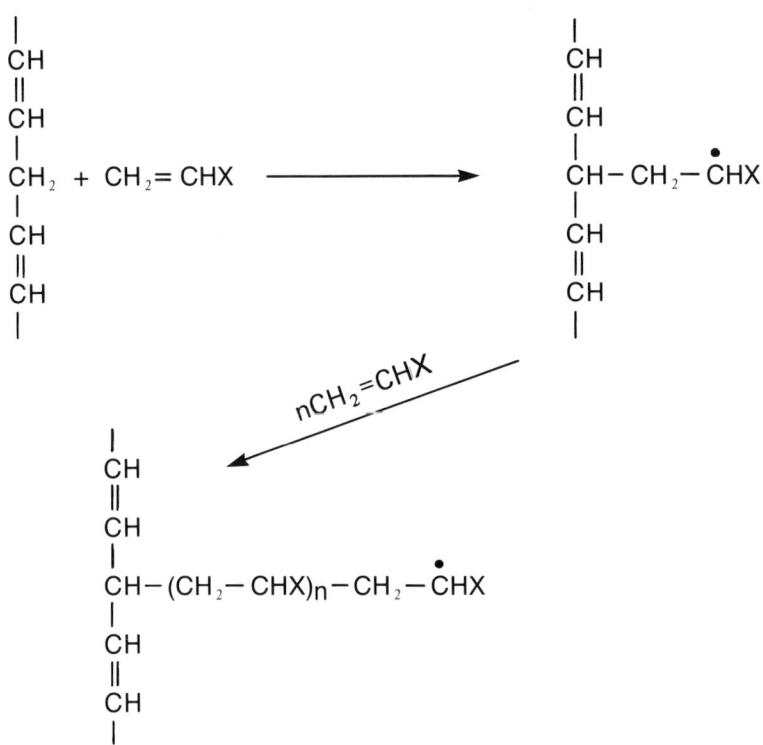

In practice the vinyl chain comprises 4−5 monomer units and a 3-D network can be built up by links between neighbouring fatty acids which may be part of the same or another alkyd resin molecule.

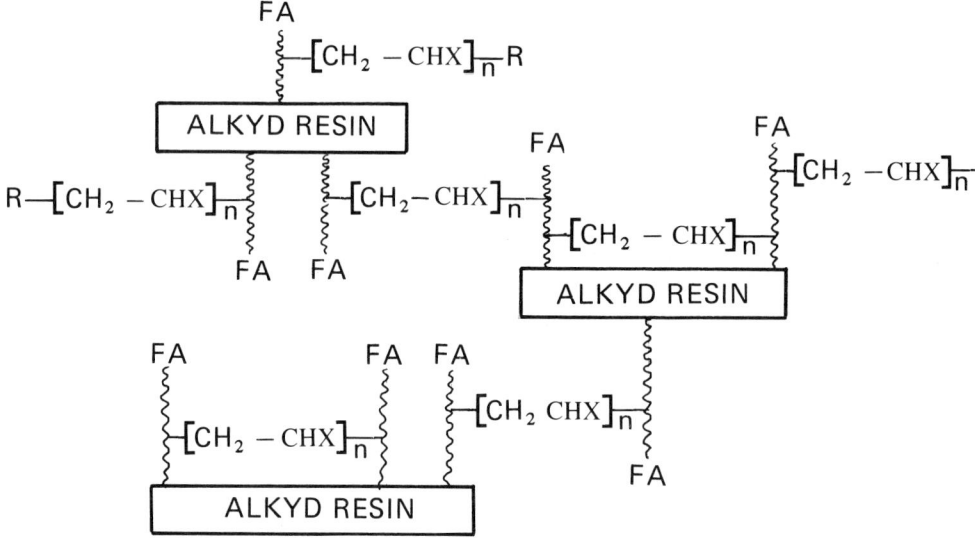

Where n depends on process conditions but is normally 4 to 5 and R is H or chain termination species.

Homopolymerisation also occurs and the resulting resin is a mixture of unmodified alkyd, copolymer and homopolymer.

With styrene and vinyl toluene a Diels Alder adduction reaction can also occur between the monomer and the conjugated fatty acid. This reaction occurs in the absence of a free radical initiator at temperatures of about 80°C but is of little use in forming a polymer chain. At higher temperatures (about 130°C) and in the presence of a free radical initiator the favoured reaction is copolymerisation.

When non-conjugated fatty acids are used, very little copolymerisation occurs with the vinyl monomer. The dominant reaction is homopolymerisation although some graft copolymer is formed at the sites of the fatty acid unsaturation.

Polymerisation of the oil also occurs to some extent, promoted by the presence of free radical initiators. Thus a vinylated alkyd, based on a non-conjugated fatty acid, results in a mixture of homopolymer, graft copolymer and unchanged alkyd resin.

For such a system to be homogeneous, the homopolymer must have a low molecular weight and there must be enough graft copolymer present to act as a solubilising bridge between the alkyd and homopolymer.

Various methods and techniques are reported for the promotion of an increased level of graft copolymer at the expense of the homopolymer. These include the use of unsaturated acids (e.g. maleic) in the alkyd formulation, and the incorporation of monomers with a high copolymerisation potential (e.g. acrylonitrile). However, the film properties of resins based on non-conjugated acids are not as good as those given by conjugated fatty acids. As a result most commercial resins are based on conjugated fatty acids. They employ non-conjugated fatty acids only in blends with conjugated fatty acids, to reduce raw material costs.

All the vinyl monomers commonly encountered in addition polymerisation can be copolymerised with alkyd resins, assuming there are sufficient sites available on the fatty acid.

Most modern vinylated alkyds are based on styrene or styrene, methyl methacrylate blends, although some resins based on vinyl toluene are still commercially available.

Vinyl toluene based systems give softer films than their styrene analogues and are not suitable for use in odour sensitive applications because of the high odour conferred, even by small traces of residual unreacted monomer. Film clarity is often poor with this type of resin and separation of vinyl toluene homopolymer often occurs on long term storage. Vinyl toluene has the advantage of being relatively inexpensive and for this reason is still used in vinylated alkyds intended for selected applications. The use of methyl methacrylate, styrene monomer combinations result in improved solvent resistance and hardness. However, the use of monomer combinations can result in homopolymers being formed by both monomer species and unless this is carefully controlled, preferential pigmentation with separation occurring on long term storage may occur. This can usually be countered when formulating the compounded coating system. Recently the monomer p-methyl styrene has been introduced as a replacement for vinyl toluene in rapid drying industrial finishes.

Where there are insufficient sites for copolymerisation, high levels of homopolymer result, leading to an incompatibility haze, which effects the gloss of the film. The film properties are also adversely effected by high levels of homopolymer. In most commercial systems the level of modification lies between 25% and 45% of the total resin content and alkyds with short or medium oil lengths are employed.

One technique, occasionally used to prepare vinylated alkyds is to copolymerise the oil or fatty acid and then use the copolymer to form the alkyd resin. This technique is also used to prepare vinyl modified glyceride oils with improved drying and tack-free properties.

The oil is heated to ca 130–150°C in the presence of the monomer and a free radical initiator, usually di-tertiary butyl peroxide or benzoyl peroxide. Chain transfer agents may also be used to control the molecular weight of the homopolymer fraction.

More usually vinylated alkyds are prepared by copolymerisation of monomer with a specifically designed alkyd base.

The alkyd is formulated by using a conjugated oil or fatty acid or blend of conjugated and non-conjugated oils or fatty acids. Oil lengths are typically 35–45%. The molecular weight of the alkyd is usually lower than for unmodified types, the functionality being typically in the range f = 1.8 to 1.9.

The most commonly employed polyol is glycerol or, where improved film performance is required, glycerol, pentaerythritol blends.

Where dehydrated castor oil is used as the conjugated oil, dehydration is often carried out in situ, using raw castor oil.

Polycondensation is carried out to meet a very narrow viscosity range. Deviation from the designated molecular weight or molecular weight distribution, can result in a sub-standard product on vinylation. The conditions under which vinylation is carried out can be altered to allow for some variation in the base alkyd, but it is the viscosity (molecular weight) of the base alkyd which determines the final viscosity of a correctly processed vinylated alkyd.

Two techniques are commonly practised for vinylation.

i) Where the level of modification is low (vinyl monomer content low), or where a wide molecular weight distribution is required, the base alkyd (diluted in solvent), the vinyl monomer and the initiator (with any chain transfer agent) are charged together to the reactor. The charge is heated to reaction temperature (normally 130–150°C) and the conversion of the monomer monitored by determination of non-volatile content. The reaction is continued until the non-volatile content is similar to the calculated theoretical. All the monomer has then been polymerised.

   The resin is cooled and diluted in the required solvent. The presence of residual initiator can cause polymerisation of the unsaturated oil residues and hence cause instability in the final product. It is normal practice therefore to restrict the amount of initiator added initially. As the initiator is used up, additional initiator is added in aliquots or 'spikes' throughout the course of the polymerisation.

ii) Where a narrower molecular weight distribution is required, the drip-feed process is used. This process has the additional advantage of producing lower levels of homopolymer (for the same number of copolymerisable sites).

   The base alkyd, together with solvent, is charged to the reactor and heated to the required reaction temperature. The vinyl monomer, premixed with initiator, is fed into the reactor over a period of several hours while maintaining the reactor at reaction temperature. With the correct choice of initiator and reaction temperature, copolymerisation will occur throughout the time the premix is added, with little build up of unreacted monomer. Polymerisation of the monomer is usually virtually complete within two hours of completion of the premix addition. Spikes of initiator are used to complete the polymerisation if required. The polymerisation is continued until complete conversion of the monomer is obtained or controlled for a particular viscosity.

The presence of residual free monomer in the resin results in substandard performance properties. The viscosity of the base alkyd influences the viscosity of the final vinylated alkyd. The base alkyd is thus formulated with this in mind.

Small variations in viscosity can be obtained by the use of all the techniques common to addition polymerisation. However, the use of high levels of initiator can lead to oil polymerisation and unstable products. The use of high levels of chain transfer agent, result in very low molecular weight homopolymer being formed. This will have adverse effects on the film properties of the resin, durability in particular, being effected.

The Uses of Vinylated Alkyds

The average molecular weight of vinylated alkyds is much higher than that of unmodified alkyds. As a result these resins form 'tack free' films very rapidly. This initial tack free stage results from solvent evaporation. Through-drying (oxidation of the film) takes much longer and this is due to the reduced levels of unsaturation in the alkyd, resulting from copolymerisation.

Vinylated alkyds based on vinyl toluene or styrene are used in industrial coatings where rapid air-drying is required. The resulting coating combines good gloss retention and recoatability with excellent durability and fast air drying and can be used to formulate machinery enamels.

Styrene methyl methacrylate modified alkyds show improved hardness over vinyl toluene and are frequently used in roller coating enamels. These resins show excellent deep draw, adhesion and tooling properties and may be force dried at 150−160°C.

Vinylated Alkyd Formulations

DCO, Styrenated Alkyd

A formulation for a vinylated alkyd resin involves, firstly, the formulation and manufacture of a long oil length alkyd as a base material.

Dehydrated castor oil	39.30
Glycerol	11.47
Phthalic anhydride	19.96
Xylene (reflux solvent)	5.14
Xylene (diluent)	24.13
	100.00
Lithium hydroxide (0.009 alcoholysis catalyst)	

PROCESS

In manufacture the dehydrated castor oil and glycerol are heated together with lithium hydroxide catalyst to 245°C for the alcoholysis reaction. Alcoholysis is deemed to be complete when a tolerance of 1 part sample to 3 parts methanol is obtained at 25°C. The reactants are cooled to 180°C and phthalic anhydride and xylene (reflux solvent) are added. Polycondensation is carried out at 210°C using the solvent process technique and the reaction is quenched when a viscosity (determined on a 70% solution of the resin in xylene) of 6–9 poises and an acid value of below 12 mgKOH/g are obtained. The solvent content of the reactor at 210°C is approximately 7% and this must be taken into consideration when diluting the sample prior to viscosity determination.

FINAL RESIN CONSTANTS

Colour	4 Gardner (maximum)
Viscosity	6–9 poises (Gardner tubes U–V)
Acid value	12 mgKOH/g (maximum)
Non-volatile content	69–71%

As a second stage, a styrene modified alkyd can be manufactured through the formulation.

Base alkyd (at 70% non-volatile content)	54.86
Xylene	20.19
Styrene	21.24
t-dodecyl mercaptan	0.15
Di-tertiary butyl peroxide (1)	0.18
Di-tertiary butyl peroxide (2)	0.03
Xylene (diluent)	3.35
	100.00

The method of manufacture can be outlined as follows: The styrene and t-dodecyl mercaptan are mixed in a premix tank and the base alkyd and xylene are heated together to 140°C. At this temperature the pre-mixed styrene and t-dodecyl mercaptan (chain transfers agent) are added to the reactor at a uniform rate over 4 hours. Simultaneously di-tertiary butyl peroxide (initiator) is added to the reactor from a separate container over the same period of time as the premixed monomers.

It is important that the initiator is kept separate from the premixed styrene and chain transfer agent to avoid the chance of premature polymerisation occurring in the premix tank.

The temperature is maintained at 140°C throughout the addition period. When the addition is complete the reactants are maintained at 140°C until conversion is complete.

The course of the reaction is monitored by determination of the non-volatile content and the reaction is complete when a constant non-volatile content of 62% is obtained. If the non-volatile content is constant, but less than 61.5% additional 'spikes' of initiator (di-tertiary butyl peroxide (2)) should be added. Initially 50% of the di-tertiary butyl peroxide (2) should be added and the non-volatile content of the reactants determined one hour after the 'spike' has been added. If there has been an increase in non-volatile content the remainder of the di-tertiary butyl peroxide (2) should be added and the reactants held to achieve a constant non-volatile content of at least 61.6%. The reactants are cooled and diluted with xylene.

FINAL RESIN CONSTANTS

Colour	4 Gardner (maximum)
Viscosity at 25°C	12 – 20 poises (Gardner tubes X – to Z –)
Acid value	below 10 mg/KOH/g
Non-volatile content	59 – 61%

APPLICATIONS

This resin is used for fast dry decorative enamels, for exterior and interior use. It has excellent durability and coating and film forming properties.

Short Oil Length Castor Oil Vinylated Alkyd

A formulation for a short oil length castor oil vinylated alkyd involves firstly the manufacture of the base alkyd.

Raw castor oil	39.80
Glycerol	3.93
Pentaerythritol	4.31
Phthalic anhydride	13.50
Solvesso 100 (diluent)	38.46
	100.00

The recommended process is that the reactants are heated together to 270°C and held at this temperature, while dehydration of the castor oil proceeds. An inert gas sweep should be maintained to remove water formed during the reaction to ensure that air is excluded from the reactor.

This viscosity falls as dehydration proceeds and once a minimum viscosity has been recorded and the viscosity begins to increase the reactor is cooled to 200°C. The polycondensation is completed at this lower temperature. Once an acid value of 10−15 mgKOH/g and a viscosity (60% in solvesso 100) of 1.5−2.0 poise have been achieved the reaction is quenched by cooling below 180°C and adding diluent.

FINAL RESIN CONSTANTS

Colour	8 Gardner
Viscosity at 25°C	1.5−2.0 poise (Gardner tubes F + to H)
Acid value	10−15 mgKOH/g
Non-volatile content	59−61%

Secondly the preparation of a short oil length castor oil vinylated alkyd follows the formulation.

Base alkyd (as above) 60% (non-volatile content)	57.03
Solvesso 100	13.65
Styrene	10.24
Methyl methacrylate	10.24
Di-tertiary butyl peroxide (1)	0.20
Di-tertiary butyl peroxide (2)	0.05
Di-tertiary butyl peroxide (3)	0.05
Solvesso 100 (diluent)	8.54
	100.00

The process involves the premixing of the styrene and methyl methacrylate in a premix tank. Then the base alkyd and solvent are heated to 135°C and the premix is added to the reactor over two hours at a uniform rate, maintaining the temperature at 135°C. At the same time the di-tertiary butyl peroxide initiator is added to the reactor at a rate which ensures that premix and initiator additions are completed at the same time. Once the premix and initiator have been added, the reactants are held at 135°C for two hours and then di-tertiary butyl peroxide (2) and (3) are added at two hourly intervals. The course of the reaction is followed by monitoring the non-volatile content and the reactants are cooled to below 100°C and diluent added when a constant non-volatile content of 60.0% is obtained. The non-volatile content is determined on 1g of sample in an oven at 150°C for 30 minutes.

FINAL RESIN CONSTANTS

Colour	6 Gardner
Viscosity	22 – 27 poises (Gardner tubes ($Z - Z_1$))
Acid value	6 – 9 mg/KOHg
Non-volatile content	54 – 56%

URETHANE MODIFIED ALKYD RESINS

Alkyd resins are frequently modified by reaction with species containing isocyanate groups, to give resins with improved film hardness and durability.

Isocyanates are highly reactive groups, which readily combine at low temperatures with alcohols to form urethane links. The reaction is an insertion reaction and no volatile reaction products are produced.

$$R - NCO \quad + \quad HO - R^1 \quad \rightarrow \quad R - NH - \overset{\displaystyle O}{\overset{\displaystyle \|}{C}} - OR^1$$

Isocyanate   Alcohol   Urethane

By using di-isocyanates to react with the residual hydroxyl groups of an alkyd resin, a high molecular weight three dimensional network can be obtained by joining the alkyd chains together via urethane links.

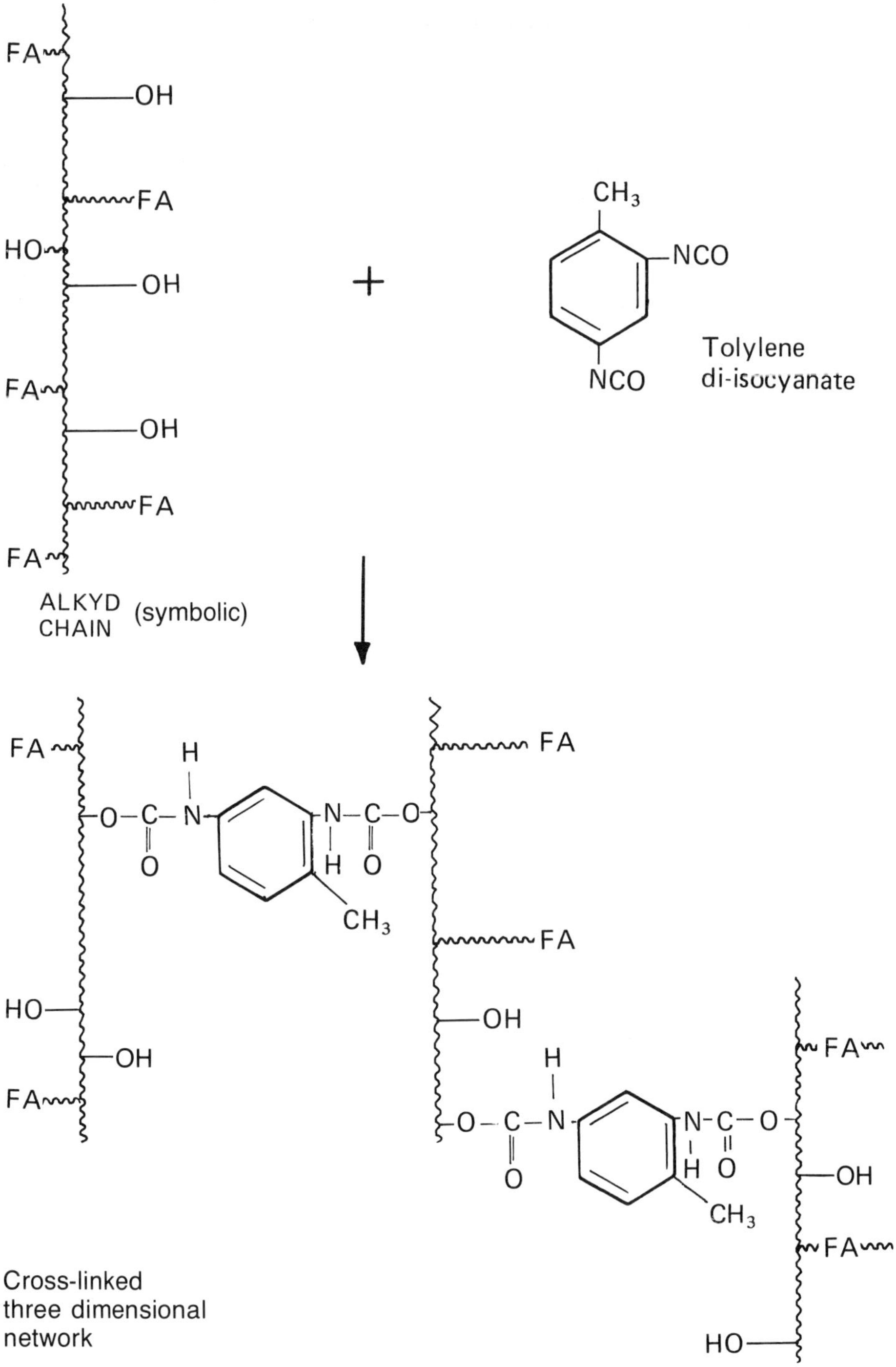
FA
OH
FA
HO
OH
FA
OH
FA
FA
ALKYD CHAIN (symbolic)
+
CH3
NCO
NCO
Tolylene di-isocyanate
FA
H
O–C–N
O
N–C–O
H O
CH3
FA
FA
HO
OH
FA
OH
H
O–C–N
O
N–C–O
H O
CH3
FA
OH
FA
HO
Cross-linked three dimensional network

A urethane alkyd may be regarded as an alkyd in which some of the di-acid has been replaced by di-isocyanate. Isocyanates are highly reactive and will react with many substances including water. Care must be taken when manufacturing urethane modified alkyds to ensure that the reactor is dry and free from contamination. It must also be ensured that the raw materials used are of a high purity, otherwise unwanted side reactions can occur. Some of these could lead to gelation. Some of the more pertinent side reactions that can occur are detailed below.

i) Reactions in presence of water

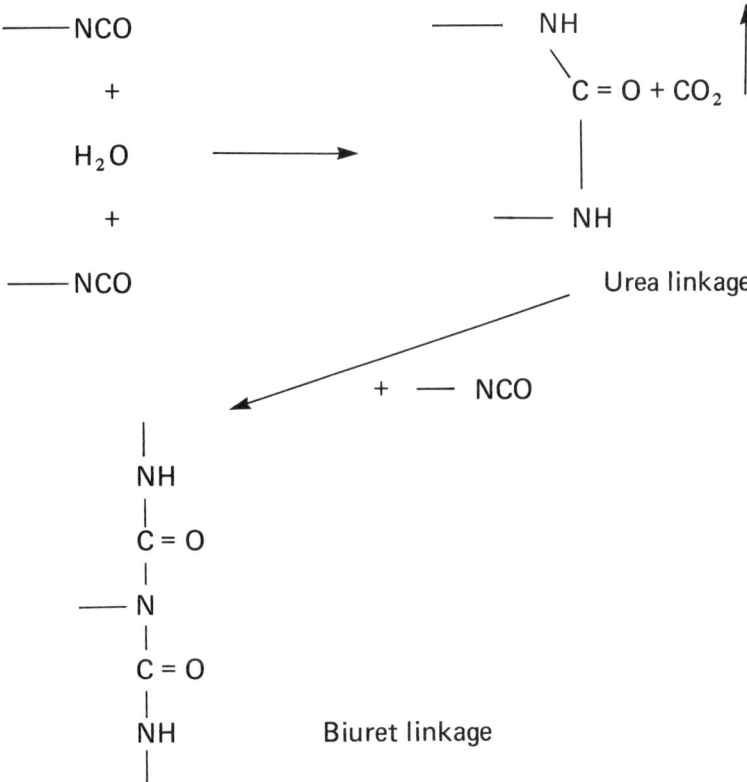

ii) Reactions in presence of excess — NCO

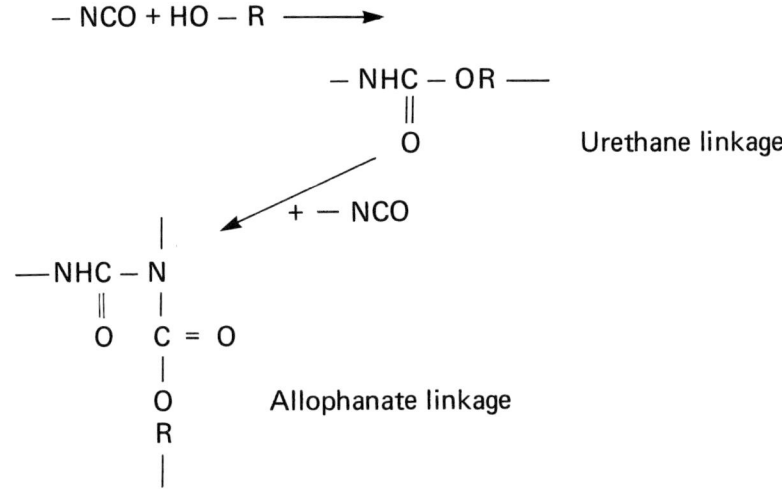

In the presence of equimolar amounts of urethane and urea the reaction will follow the Biuret course since this is the favoured reaction. It can thus be seen that to avoid the formation of highly cross-linked products, or even gelation, it is necessary to ensure that water is excluded and that efficient agitation is employed to avoid localised concentrations of isocyanate being formed during manufacture.

Tolylene di-isocyanate (T.D.I.) is the preferred isocyanate for urethane alkyd manufacture, despite its high toxicity and the handling precautions which this necessitates. Tolylene di-isocyanate is supplied commercially as 80/20 T.D.I. which is a mixture of 80% of the 2:4 isomer and 20% of the 2:6 isomer.

2 : 4 isomer   2 : 6 isomer

The reactivity of the isocyanate group, in the para position, with hydroxyl groups is about eight times that of the isocyanate group in the ortho position. This allows more control of the reaction during manufacture.

Because of its toxicity, T.D.I. is normally delivered to the reactor via closed systems, pressurised with inert gas (usually nitrogen). It is important that the nitrogen is completely dry and that all water vapour is removed normally by passing the nitrogen through drying towers before it contacts the T.D.I. Similarly the reactor should be swept with a dry nitrogen purge.

Urethane Alkyd Formulations

FORMULATION

Soya bean oil fatty acid	34.01
Trimethylol propane	13.54
Calcium naphthenate (4% Ca)	0.12
Tolylene di-isocyanate	13.25
White spirit (diluent)	39.08
	100.00

PROCESS

The process for its manufacture involves heating the fatty acid and tri-methylol propane to 220°C under an inert gas sweep. The reactants are held at 220°C until an acid value below 5 mgKOH/g is obtained. Then they are cooled to 100°C. At 100°C the calcium naphthenate catalyst and tolylene di-isocyanate are added. There is a vigorous exotherm and the temperature must be maintained below 135°C. The reactants are held at 125−135°C under sealed reactor conditions, until an isocyanate value of below 1 mgKOH/g is obtained. The reactants are then cooled and diluted with white spirit.

FINAL RESIN CONSTANTS

Colour	8 Gardner (maximum)
Viscosity at 25°C	36−46 poises (Gardner tubes $Z_2 - Z_3$)
Acid value	4 mgKOH/g (maximum)
Non-volatile content	58−60%

APPLICATIONS

This type of resin finds applications in clear and pigmented one pack polyurethane finishes, either on its own, or in combination with thixotropic or unmodified drying alkyds and the finishes have rapid drying properties and excellent chemical and oil resistance. This type of resin frequently finds use in marine varnishes.

FORMULATION

Soya bean oil fatty acid	14.75
Dehydrated castor oil fatty acid	14.75
Pentaerythritol	9.99
Phthalic anhydride	4.90
Calcium naphthenate (4% Ca)	0.07
Tolylene di-isocyanate	6.76
White spirit	48.78
	100.00

Its manufacture involves the heating of the fatty acids, pentaerythritol and phthalic anhydride together to 240°C under an inert gas sweep. Polycondensation is carried out, until an acid value below 5 mgKOH/g is recorded. At this point the reactor is cooled to 100°C.

At 100°C, calcium naphthenate is added and the tolylene di-isocyanate is fed to the reactor over 30 minutes. There is a vigorous exotherm, which must be controlled so that the reactor temperature is maintained at 125–135°C. The course of urethane formation is followed by the determination of isocyanate value. Once the isocyanate value is below 1 mgKOH/g, white spirit is added and the reactants cooled.

FINAL RESIN CONSTANTS

Colour	6 Gardner (maximum)
Viscosity at 25°C	2.5–3.0 poises (Gardner tubes H to L)
Acid value	3 mgKOH/g (maximum)
Non-volatile content	49–51%

APPLICATIONS

This resin is designed for use in clear or pigmented systems as a one-pack cure urethane finish. It may be used on its own, or in blends with other drying alkyds. It confers excellent hardness, gloss and rapid drying and has excellent adhesion and good chemical resistance.

Urethane alkyds have better all round performance properties than urethane oils, although the latter are frequently used in blends with unmodified alkyds.

Film formation occurs, via the unsaturation of the fatty acids as with unmodified alkyds, but the high molecular weight of the urethane alkyd gives increased drying speed.

The dried film exhibits superior hardness, much improved durability and alkali resistance.

As a consequence these resins are frequently used in marine varnishes and metal primers. They are also used as blends with unmodified alkyds in air-drying decorative paints. However, films formed from alkyds containing high levels of urethane modification, show an increased tendency to yellow. Urethane alkyds and oils also find application in printing ink formulations, but many applications are restricted due to the incorporation of T.D.I. and the S.B.P.I.M. guidleines.

POLYAMIDE MODIFIED ALKYD RESINS

The rheology of an alkyd resin, can be modified by reaction with specially designated polyamide resins (see chapter on polyamides). The resultant product is a jelly-like structured material which breaks down under shear (e.g. brushing or stirring) to a free flowing liquid. Once the shear is removed the resin re-sets to a jelly. The degree of structure and the rate at which the resin re-sets is dependent on the level and type of polyamide used. These resins form the basis of 'non-drip' or 'thixotropic' oil paints.

The reaction between the alkyd resin and the polyamide is an ester interchange reaction.

Initially the alkyd and polyamide species are mutually insoluble. As interchange proceeds, polyamide molecules are reacted onto some of the alkyd resin chains. This modified alkyd acts as a solubilising bridge between unmodified alkyd resin and unreacted polyamide. The level of inter-action obtained between the two species, determines the nature of the final product. Where the ratio of modified alkyd to unreacted polyamide and alkyd is low, the degree of structure exhibited by the product is high (rigid gel). The mutual solubility of the system is however, poor and separation, particularly after pigmentation, can occur. Where the ratio of modified alkyd to unreacted polyamide and alkyd is high, good mutual solubility results, but the degree of structure exhibited by the product is low (soft gel). Precise control of the polyamide alkyd interchange is required and most commercial resin manufacturers have developed their own special techniques which allow accurate assessment and control of the degree of reaction.

Polyamide modified alkyds are visco-elastic in their behaviour to stress. They are predominantly viscous with elastic overtones.

Viscous flow is irreversible bulk deformation of the polymer. It is time dependent and is associated with the irreversible slippage of molecules or chains past each other.

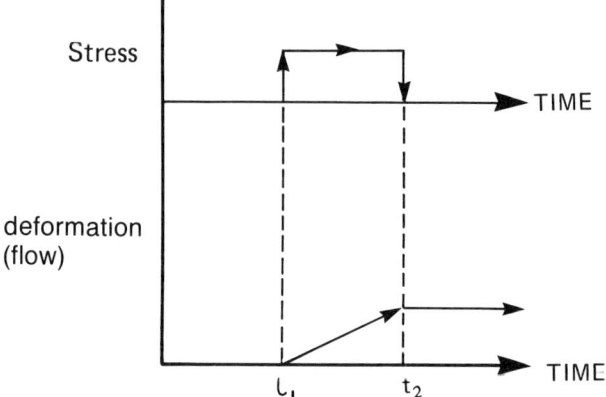

If a fixed stress is applied, to a body exhibiting true viscous flow, at time t_1 a deformation is obtained which increases until t_2 when the stress is released. The deformation then remains constant with time. There is no recovery and deformation is not instantaneous.

On the other hand elastic flow is reversible and time dependent. Here the motion of the polymer chain segments is restricted by chain entanglement. This type of flow probably only involves bond stretching and bond angle deformation.

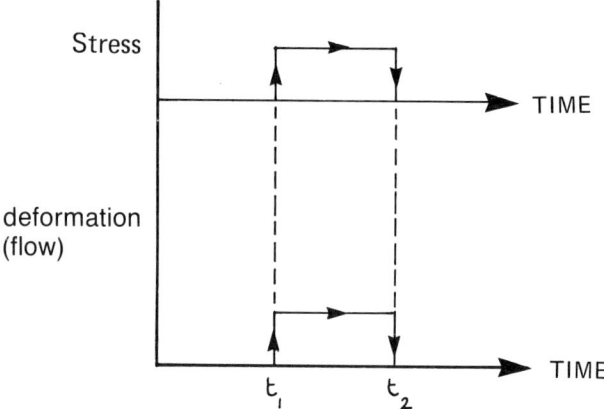

If a stress is applied at time t_1, a corresponding strain occurs instantly and remains constant until the stress is removed at time t_2, when the strain instantly returns to zero. There is instantaneous deformation, which is completely reversible when the stress is removed.

Only Newtonian fluids exhibit ideal viscous flow and only Hookean solids exhibit ideal elastic flow.

Polymers are non-Newtonian in their behaviour and may be divided into two main classes.

1. **Time independent fluids:** are those where the shear stress, as measured by the apparent viscosity (ηa), is a function of the shear rate. ($\overset{\cdot}{\gamma}$).

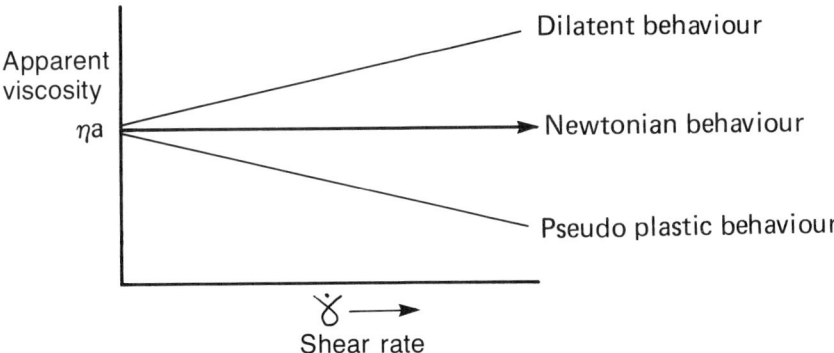

Dilatent materials exhibit an increase in apparent viscosity, with increase in shear rate. Pseudo plastic materials exhibit a decrease in apparent viscosity with increase in shear rate.

Newtonian materials are characterised by an apparent viscosity, which is independent of shear rate.

2. **Time dependent fluids:** are those where the shear stress/shear rate relationship depends upon the shear time.

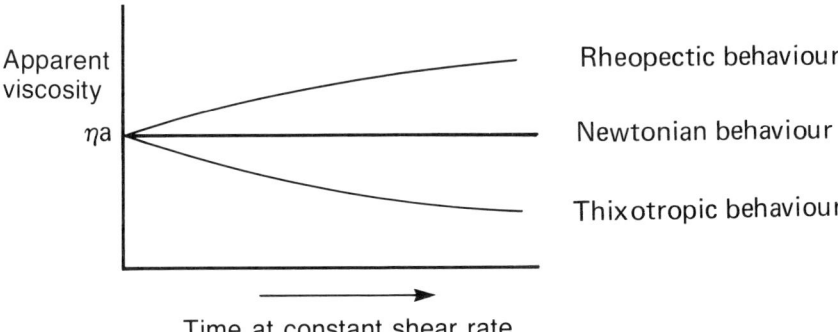

Here there are two classes of behaviour. Those materials which shear thicken with time are called Rheopectic and those materials that show a decrease in apparent viscosity with time at constant shear rate, are said to be thixotropic in their behaviour.

Polyamide modified alkyds exhibit thixotropic behaviour. If subjected to a constant shear, the viscosity falls with time to a minimum value, after which there is no further decrease in viscosity.

A plot of strain against shear rate, shows that the deformation is directly related to shear rate and when the shear is removed the deformation returns after a time to zero, i.e. the viscosity returns to its original value, over a period of time, once shear is removed.

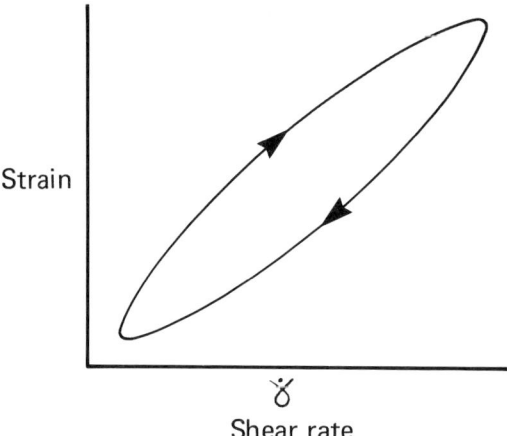

Strain

$\overset{\smallsmile}{\gamma}$

Shear rate

Polyamide modified alkyds behave in this way, due to the complex chain entanglements that occur between the two mutually insoluble species (polyamide and unmodified alkyd) and the solubilising species (polyamide modified alkyd).

Where there is no unreacted polyamide and the system consists of alkyd and modified alkyd, little or no thixotropic structure is observed.

The application of shear (e.g. brushing of a polyamide modified alkyd) has the effect of separating the chain entangled species (although high shear and temperatures above 60°C are required to obtain complete disentanglement) with a consequent decrease in viscosity enabling the resin to flow. Once the shear is removed the resin begins to restructure, as the stressed polymer chains relax and assume their previous entangled state.

When correctly formulated, a paint can be produced which will flow evenly on brush application but 'set-up' rapidly after application.

The Uses of Polyamide Modified Alkyd Resins

These resins are used exclusively in air-drying decorative paints, where their rheological properties make them attractive to D.I.Y. home decorators.

They are frequently used as blends, with unmodified alkyds or urethane modified alkyds, to impart structure.

Care must be taken when compounding a paint from these resins, since separation can occur in blends with other types of alkyd and each resin manufacturer usually supplies details of recommended blend ratios. Most manufacturers also recommend that low shear conditions should be employed when the paint making operation is likely to exceed temperatures of ca 40°C. This is to avoid a permanent reduction in the degree of structure.

The structure of these systems depends on the solubility differences of the polyamide and alkyd. The addition of solvents which improve the solubility of the system can lead to loss of structure. Therefore the addition of polar solvents such as alcohols, or aromatic solvents such as xylene must not be used (although small amounts of xylene used as process solvent may be tolerated).

Formulation for a Thixotropic Alkyd

FORMULATION

Soya bean oil	29.45
Pentaerythritol	6.43
Phthalic anhydride	16.94
Fumaric acid	0.19
Xylene (reflux solvent)	1.92
Polyamide (e.g. Wolfamid 500)	2.30
White spirit	48.33
	100.00

Lithium hydroxide (0.008 alcoholysis catalyst)

PROCESS

Its manufacture involves:

1. **Alcoholysis:** soya bean oil and pentaerythritol are alcoholysed at 245°C, using lithium hydroxide as alcoholysis catalyst. Once a tolerance of 1 part sample to 3 parts methanol (at 25°C) is obtained the reactants are cooled to below 180°C and phthalic anhydride, fumaric acid and xylene are added.

2. **Polycondensation:** is carried out at 260°C under solvent reflux. Once an acid value of 16 mgKOH/g has been achieved the xylene is removed by vacuum distillation and esterification continued, using a fusion process until an acid value below 12 mgKOH/g and viscosity (70% in white spirit) of 20−25 poises at 25°C are achieved.

3. **Polyamide Reaction:** the reactants are cooled to 210°C and the polyamide added. The reactor is held at 200−210°C while the polyamide reacts with the alkyd. The course of the reaction is monitored by determination of the clarity of a 10% solution in white spirit. When the resin sample is completely clear, the reactants are cooled immediately and diluted with white spirit.

APPLICATIONS

This resin is designed for blending with unmodified alkyds to produce thixotropic decorative paints. The above formulation results in a high gel strength product. Where lower degrees of structure are required the polyamide content may be reduced or a lower molecular weight polyamide used.

FORMULARY

Vinylated Alkyd Resins

The vinylation of an alkyd resin improves its film hardness and chemical resistance as well as shortening its drying time. These resins are generally used where rapid drying performance and toughness are required.

ROAD MARKING PAINT

	Anatase titanium di-oxide	7.56
	Calcite	45.39
	Whiting	7.56
	*Vinyl toluene modified alkyd (50% non-volatile content)	31.77
Driers	Lead naphthenate (24% Pb)	0.07
	Cobalt naphthenate (6% Co)	0.03
	Methyl ethyl ketoxime	0.02
	High boiling petroleum solvent (Bpt. 150 – 170°C)	0.35
	Xylene	7.15
		100.00

*Modified Alkyd used:
Vinyl toluene modified short oil length alkyd,
typical level of modification ca 35 – 40%.

RAPID DRYING WHITE GLOSS FINISH

	Rutile titanium di-oxide	25.30
	*Styrene modified alkyd (50% non-volatile content)	65.30
	High boiling petroleum solvent (B.pt 150 – 170°C)	7.75
Driers	Lead naphthenate (24% Pb)	0.62
	Cobalt naphthenate (6% Co)	0.31
	Calcium naphthenate (4% Ca)	0.62
	Methyl ethyl ketoxime	0.10
		100.00

WHITE ROLLER COATING ENAMEL

Rutile titanium di-oxide	29.40
*S/MMA modified alkyd (55% non-volatile content)	52.00
High boiling aromatic hydrocarbon (e.g. Shellsol AB Solvesso 150)	18.60
	100.00

*Modified Alkyd used:
Short oil D.C.O. alkyd modified with a styrene/methyl methacrylate level of modification approximately 45%.

URETHANE OILS — CLEAR LACQUER

20–30% urethane modified soya bean oil/penta ester (60% non-volatile content)	91.51
Cobalt naphthenate (6% Co)	0.09
White spirit	8.40
	100.00

WHITE GLOSS FINISH

	Rutile titanium di-oxide	24.50
	20–30% urethane modified soya penta ester (60% non-volatile content)	55.70
Drier	Cobalt naphthenate (6% Co)	0.20
	Lead naphthenate (24% Pb)	0.40
	Methyl ethyl ketoxime	0.08
	White spirit	19.12
		100.00

Both the above systems will have good chemical resistance and excellent drying and adhesion properties.

Urethane Alkyds

These resins may be used as sole binders for paints and printing inks or they may be blended with unmodified alkyd resins to improve film properties and drying time or with thixotropic alkyds to give films with improved drying time.

When used as a blend in conjunction with other non-urethane modified alkyds the urethane modified alkyd normally comprises 20−50% of the total binder content.

RED GLOSS FINISH

	Permanent Red R	9.10
	*13% urethane modified alkyd (60% non-volatile content)	41.30
	Long oil oxidising alkyd (60% non-volatile content)	38.60
Driers	Calcium naphthenate (5% Ca)	0.91
	Lead naphthenate (24% Pb)	0.95
	Cobalt naphthenate (6% Co)	0.38
	Methyl ethyl ketoxime	0.13
	White spirit	8.63
		100.00

*Urethane alkyd used:
12−15% urethane modified long oil oxidising alkyd

GREEN GLOSS URETHANE ALKYD FINISH

	Middle chrome green	17.50
	13% urethane modified alkyd (60% non-volatile content)	23.90
	Long oil, tall oil or soya bean oil alkyd (70% non-volatile content)	39.11
Driers	Calcium naphthenate (5% Ca)	0.90
	Lead naphthenate (24% Pb)	0.90
	Cobalt naphthenate (6% Co)	0.37
	Methyl ethyl ketoxime	0.13
	White spirit	17.19
		100.00

WHITE GLOSS FINISH

	Rutile titanium di-oxide	27.60
	13% urethane modified alkyd (60% non-volatile content)	20.10
	Long oil, soya penta alkyd (60% non-volatile content)	42.80
Driers	Calcium naphthenate (5% Ca)	0.74
	Lead naphthenate (24% Pb)	0.76
	Cobalt naphthenate (6% Co)	0.31
	Methyl ethyl ketoxime	0.11
	White spirit	7.58
		100.00

CLEAR FLOOR VARNISH

	15% urethane modified alkyd	89.30
Driers	Cobalt naphthenate (6% Co)	0.25
	Lead naphthenate (24% Pb)	0.61
	Methyl ethyl ketoxime	0.15
	White spirit	9.69
		100.00

Thixotropic Alkyds

Thixotropic alkyds may be used on their own or in blends with urethane modified or un-modified alkyd resins. They can be used in the field of air drying non-drip decorative finishes.

WHITE AIRLESS SPRAY URETHANE GLOSS FINISH

	Rutile titanium di-oxide	26.00
	Soya lecithin	0.26
	13.5% urethane modified long oil soya alkyd (55% non-volatile content)	46.00
	Thixotropic long oil soya alkyd (50% non-volatile content)	13.60
Driers {	Cobalt naphthenate (6% Co)	0.30
	Lead naphthenate (24% Pb)	0.72
	Methyl ethyl ketoxime	0.18
	Solvent SBP6 (ex-Shell Chemicals)	6.47
	High flash point white spirit	6.47
		100.00

WHITE GLOSS FINISH

	Rutile titanium di-oxide	26.70
	Long oil soya alkyd (60% non-volatile content)	35.40
	Thixotropic long oil soya alkyd (50% non-volatile content)	28.40
Driers {	Cobalt naphthenate (6% Co)	0.31
	Lead naphthenate (24% Pb)	0.78
	Calcium naphthenate (5% Ca)	0.89
	Methyl ethyl ketoxime	0.18
	White spirit	7.34
		100.00

INDEX

222